The Internet of Things—Current Trends, Applications and Future Challenges

The Internet of Things—Current Trends, Applications and Future Challenges

Editor

Sergio Correia

MDPI

Basel • Beijing • Wuhan • Barcelona • Belgrade • Novi Sad • Cluj • Manchester

Editor
Sergio Correia
Portalegre Polytechnic
University
Portalegre
Portugal

Editorial Office
MDPI AG
Grosspeteranlage 5
4052 Basel, Switzerland

This is a reprint of articles from the Special Issue published online in the open access journal *Computers* (ISSN 2073-431X) (available at: www.mdpi.com/journal/computers/special_issues/Iot_2022).

For citation purposes, cite each article independently as indicated on the article page online and as indicated below:

Lastname, A.A.; Lastname, B.B. Article Title. *Journal Name* **Year**, *Volume Number*, Page Range.

ISBN 978-3-7258-1622-4 (Hbk)
ISBN 978-3-7258-1621-7 (PDF)
doi.org/10.3390/books978-3-7258-1621-7

Contents

About the Editor . . . . **vii**

Preface . . . . **ix**

Andrzej Ożadowicz
Generic IoT for Smart Buildings and Field-Level Automation—Challenges, Threats, Approaches, and Solutions
Reprinted from: *Computers* **2024**, *13*, 45, doi:10.3390/computers13020045 . . . . **1**

Konstantinos Charmanas, Konstantinos Georgiou, Nikolaos Mittas and Lefteris Angelis
Classifying the Main Technology Clusters and Assignees of Home Automation Networks Using Patent Classifications
Reprinted from: *Computers* **2023**, *12*, 211, doi:10.3390/computers12100211 . . . . **33**

Yarob Abdullah and Zeinab Movahedi
QoS-Aware and Energy Data Management in Industrial IoT
Reprinted from: *Computers* **2023**, *12*, 203, doi:10.3390/computers12100203 . . . . **51**

Omar Banimelhem and Fidaa Al-Quran
Rendezvous Based Adaptive Path Construction for Mobile Sink in WSNs Using Fuzzy Logic
Reprinted from: *Computers* **2023**, *12*, 66, doi:10.3390/computers12030066 . . . . **65**

Mandli Rami Reddy, M. L. Ravi Chandra, P. Venkatramana and Ravilla Dilli
Energy-Efficient Cluster Head Selection in Wireless Sensor Networks Using an Improved Grey Wolf Optimization Algorithm
Reprinted from: *Computers* **2023**, *12*, 35, doi:10.3390/computers12020035 . . . . **80**

Tanweer Alam
Blockchain-Based Internet of Things: Review, Current Trends, Applications, and Future Challenges
Reprinted from: *Computers* **2023**, *12*, 6, doi:10.3390/computers12010006 . . . . **97**

Zlatin Zlatev, Tsvetelina Georgieva, Apostol Todorov and Vanya Stoykova
Energy Efficiency of IoT Networks for Environmental Parameters of Bulgarian Cities
Reprinted from: *Computers* **2022**, *11*, 81, doi:10.3390/computers11050081 . . . . **127**

Evgeny Kalinin, Danila Belyakov, Dmitry Bragin and Konev Anton
IoT Security Mechanisms in the Example of BLE
Reprinted from: *Computers* **2021**, *10*, 162, doi:10.3390/computers10120162 . . . . **140**

Yahya Tashtoush, Israa Haj-Mahmoud, Omar Darwish, Majdi Maabreh, Belal Alsinglawi, Mahmoud Elkhodr and Nasser Alsaedi
Enhancing Robots Navigation in Internet of Things Indoor Systems
Reprinted from: *Computers* **2021**, *10*, 153, doi:10.3390/computers10110153 . . . . **149**

Sotiroula Theodosi and Iolie Nicolaidou
Affecting Young Children's Knowledge, Attitudes, and Behaviors for Ultraviolet Radiation Protection through the Internet of Things: A Quasi-Experimental Study
Reprinted from: *Computers* **2021**, *10*, 137, doi:10.3390/computers10110137 . . . . **173**

About the Editor

Sergio Correia

Prof. Sérgio D. Correia received his Diploma in Electrical and Computer Engineering from the University of Coimbra, Portugal, in 2000, his Master's Degree in Industrial Control and Maintenance Systems from Beira Interior University, Covilhã, Portugal, in 2010, and a Ph.D. in Electrical and Computer Engineering from the University of Coimbra, Portugal, in 2020. He is a Professor at the Portalegre Polytechnic University, Portugal, where he is the head of the Laboratory for Electronics and Instrumentation (LEI). He is a Researcher at NOVA School of Science and Technology, Center of Technology and Systems (UNINOVA-CTS), and VALORIZA - Research Center for Endogenous Resource Valorization, Portalegre Polytechnic University, Portalegre, Portugal, and has worked with several private companies for more than 20 years. His current research interests are Embedded Artificial Intelligence, Soft Computing, Embedded Systems, and Computer Architecture, topics on which he is an editor of several scientific journals, a collaborator at several International Conferences, and a visiting professor at several European universities.

Preface

The Internet of Things represents one of the most significant technological advancements of our era. It interconnects a multitude of devices and systems, thereby transforming how we interact with the world around us. The breadth and depth of its applications span various domains, offering innovative solutions and posing new challenges. This reprint explores these dimensions through a curated selection of research studies and reviews that address the multifaceted nature of IoT technologies. We are grateful to all the contributing authors for their valuable insights and pioneering research.

The opening chapter, "Generic IoT for Smart Buildings and Field-Level Automation—Challenges, Threats, Approaches, and Solutions," delves into the complexities of implementing IoT in smart buildings. Following, "Classifying the Main Technology Clusters and Assignees of Home Automation Networks Using Patent Classifications" provides a meticulous analysis of home automation networks. The reprint then transitions to industrial applications with "QoS-Aware and Energy Data Management in Industrial IoT." This chapter focuses on the critical aspects of Quality of Service (QoS) and energy management, which are essential for optimizing industrial IoT operations. Further advancing the discussion on IoT efficiency, "Rendezvous Based Adaptive Path Construction for Mobile Sink in WSNs Using Fuzzy Logic" introduces an innovative approach utilizing fuzzy logic. "Energy-Efficient Cluster Head Selection in Wireless Sensor Networks Using an Improved Grey Wolf Optimization Algorithm" presents a novel algorithm designed to improve energy efficiency in WSNs. The integration of blockchain technology with IoT is examined in "Blockchain-Based Internet of Things: Review, Current Trends, Applications, and Future Challenges." Environmental sustainability is addressed in "Energy Efficiency of IoT Networks for Environmental Parameters of Bulgarian Cities." This chapter evaluates the energy efficiency of IoT networks in monitoring and managing environmental parameters, using Bulgarian cities as case studies. Security, a paramount concern in IoT, is the focus of "IoT Security Mechanisms in the Example of BLE." This chapter discusses security mechanisms specific to Bluetooth Low Energy (BLE), providing insights into protecting IoT ecosystems from emerging threats. "Enhancing Robots Navigation in Internet of Things Indoor Systems" explores robots' navigation capabilities within IoT-enabled indoor systems. The chapter presents advancements in robotic navigation facilitated by IoT technologies, which enhance the functionality and reliability of indoor systems. Lastly, "Affecting Young Children's Knowledge, Attitudes, and Behaviors for Ultraviolet Radiation Protection through the Internet of Things: A Quasi-Experimental Study" investigates the educational impact of IoT on health.

Sergio Correia
Editor

computers MDPI

Review

Generic IoT for Smart Buildings and Field-Level Automation—Challenges, Threats, Approaches, and Solutions

Andrzej Ożadowicz

Department of Power Electronics and Energy Control Systems, Faculty of Electrical Engineering, Automatics, Computer Science and Biomedical Engineering, AGH University of Krakow, al. Mickiewicza 30, 30-059 Krakow, Poland; ozadow@agh.edu.pl; Tel.: +48-12-617-50-11

Abstract: Smart home and building systems are popular solutions that support maintaining comfort and safety and improve energy efficiency in buildings. However, dynamically developing distributed network technologies, in particular the Internet of Things (IoT), are increasingly entering the above-mentioned application areas of building automation, offering new functional possibilities. The result of these processes is the emergence of many different solutions that combine field-level and information and communications technology (ICT) networks in various configurations and architectures. New paradigms are also emerging, such as edge and fog computing, providing support for local monitoring and control networks in the implementation of advanced functions and algorithms, including machine learning and artificial intelligence mechanisms. This paper collects state-of-the-art information in these areas, providing a systematic review of the literature and case studies with an analysis of selected development trends. The author systematized this information in the context of the potential development of building automation systems. Based on the conclusions of this analysis and discussion, a framework for the development of the Generic IoT paradigm in smart home and building applications has been proposed, along with a strengths, weaknesses, opportunities, and threats (SWOT) analysis of its usability. Future works are proposed as well.

Keywords: building automation; Internet of Things; generic IoT; fieldbus; edge computing; fog computing; blockchain; machine learning; artificial intelligence; IoT assessment

Citation: Ożadowicz, A. Generic IoT for Smart Buildings and Field-Level Automation—Challenges, Threats, Approaches, and Solutions. *Computers* **2024**, *13*, 45. https://doi.org/10.3390/computers13020045

Academic Editor: Sergio Correia

Received: 22 December 2023
Revised: 21 January 2024
Accepted: 30 January 2024
Published: 3 February 2024

1. Introduction

The rapid development of Internet-of-Things (IoT) technology over the last dozen or so years has contributed to a significant increase in the diversity and heterogeneity of data-transmission networks, as well as the emergence of numerous standards, communication protocols, and approaches in network organization. However, technological progress and development in this area are inevitable, and the IoT paradigm is constantly gaining new application areas, in particular in the field of smart solutions, both industrial, commercial, and utility, dedicated directly to customers [1,2]. This diversity of applications generates the need to consider the various requirements and expectations of users and applications. For example, the implementation of IoT devices (smart nodes) in the control and monitoring network of a production line with industrial robots poses completely different challenges to network installers and integrators than in the case of identifying products and a purchasing application based on smart IoT labels or operating IoT modules with a smartphone in smart-home or smart-city installations [2–4]. Therefore, it is a difficult task to define and sanction a uniform universal generic IoT framework for all its potential application areas, although many research and engineering teams are making efforts. This paper focuses on the applications of smart-home technologies and smart-building systems, considering their inclusion in larger system structures such as local energy microgrids and smart cities [5–7].

1.1. Fieldbus Networks in Smart Homes and Buildings

Over the last thirty years, network-communication technologies and protocols for building and industrial automation systems, as well as information and communications technology (ICT) networks, have been developing independently of each other. In particular, to support various sensor and actuator modules characteristic for building automation and control systems (BACS) and building management systems (BMS) dedicated communication technologies have been developed to transmit short data packets in channels with relatively low bandwidth that are sufficient for this type of use. Physically, communication was and is implemented in most such systems through data buses with various types of communication media, most often a twisted pair and less often a power line, optical fiber, or radio channels. Three open, international standards, namely KNX [8], LonWorks [9], and BACnet [10], have been developed for BACS, dedicated to the implementation of fieldbus networks in buildings [11–14]. In addition, many manufacturers of building automation modules have developed and introduced their own, proprietary communication protocols, reserved to support the modules and devices they offer. Most of them are based on the seven-layer open systems interconnection model from the International Organization for Standardization (ISO/OSI model) providing a common basis for the purpose of systems interconnection [15–17]. Based on these technologies and protocols, distributed automation networks are built, operating in the event-based regime [18–21].

In turn, over the last ten to fifteen years, distributed systems with wireless communication of various technologies (Bluetooth, ZigBee, Z-Wave, and Wi-Fi) have become increasingly popular in the smart-home and smart-building market, especially for solutions dedicated to commercial and individual customers. In most cases, the organizational concept of such networks is based on a simplified configuration of network devices and control functions, using a smartphone, mobile devices, or Web services [22–26]. However, it should be noted that this integration in the physical and communication layer is mostly based on wireless Wi-Fi channels and the TCP/IP protocol. This is primarily due to the rapid increase in the popularity of Wi-Fi access points in private, public, commercial, and industrial buildings as an element of ensuring continuous, mobile, and remote communication with the buildings' infrastructure as well as their users/occupants. Moreover, people have used to almost constant use of smartphones and applications dedicated to them, including smart-home control and monitoring with Wi-Fi communication.

The mentioned wireless-communication technologies, such as Bluetooth Low Energy (BLE), ZigBee, or Z-Wave, tend to be used locally at the field level in home and building automation systems. This is mainly due to their short signal range and low power consumption. Therefore, they are implemented in various types of sensors (local measurement of temperature, pressure, CO_2 concentration, or light intensity) [26–31] and modules for location, user-activity tracking (beacons), and presence verification. In these applications, their short-range characteristics increase the precision of operation and can support the implementation of some building automation functions related to thermal comfort, adaptive control of heating, ventilation, and air conditioning (HVAC) systems, or dynamic regulation of room lighting [32–37]. Therefore, in relation to the generic IoT concept, they should only be considered as supporting technologies that cooperate with active nodes of KNX, LonWorks, BACnet, and ICT networks with TCP/IP protocol dedicated to IoT solutions.

These trends have opened the way to the expansion of IoT concepts and technologies as a solution not only for remote access to distributed BACS and BMS networks but also for communicating distributed modules with the TCP/IP protocol interface within such systems, at the field level as well. At the same time, however, the prospect of integrating ICT network technologies with fieldbus network technologies created the need to solve several technological and application problems [3,5,13,19,23,38,39]:

- Field-level IP protocol implementation with real-time requirements;
- Development of IoT structures for fieldbus networks;
- Assumptions for implementation of edge and fog services;
- Big-data processing within the edge and cloud computing for BACS and BMS;

- Cybersecurity and data privacy;
- Energy efficiency and energy consumption reduction for wireless modules, sensors, and actuators.

1.2. Edge and Fog Computing within Advanced Home and Building Automation Systems

The network and functional structures of modern BACS are characterized by the two most important features: (i) distribution—the possibility of installing universal microcontrollers performing simple control, monitoring, and communication functions directly in network nodes at the field level and (ii) integration—striving for mutual connection in one exchange network data control and monitoring functions for as many devices and building infrastructure subsystems as possible. Advances in digital technologies, as well as embedded systems based on System-on-a-Chip (SoC) architectures, have enabled the development of many control and monitoring devices (network nodes) that are powerful for support control and data communication functions directly at the field level, near the building infrastructure (for example, temperature and occupancy sensors, various actuators, such as valves, motors, etc.) [3,5,40]. In previous and some of the existing solutions, most of the functions of control modules, along with the data exchange between them, were implemented at the field level. However, with this approach, the desire to include an increasing number of new building infrastructure elements and devices in the network resulted in an increase in the resource load of these modules (memory, processor) and the use of communication bus bandwidth. Moreover, the prospect of developing BACS and BMS with the functions of dynamic response to changes in parameters and decision making in the implementation of energy-efficiency improvement mechanisms and transactive energy (for example, demand side management—DSM and demand side response—DSR) requires maintaining high time determinism and working in real-time mode (with minimal, deterministic data communication delays) [41–44].

To improve the responsiveness and correctness of the BACS and BMS, edge intelligence and devices have been proposed. They push processing for data-intensive, advanced control and monitoring functions away from the field-level nodes to a new edge network level, effectively handle local workloads, and make faster, more precise service decisions. Therefore, in the concept of smart home and smart building, one of the solutions turned out to be the expansion of BACS and BMS network structures with new SoC edge modules. They communicate with field-level nodes to collect data from sensors or provide signals to actuators, and, at the same time, they are responsible for handling higher-level data communication and local processing of advanced algorithms for monitoring, data acquisition, and control functions. Moreover, these edge modules, thanks to routing support and the inclusion in the TCP/IP network, also allow communication of smart-home and smart-building modules with external cloud services (databases, data analytics and visualization, cooperation with machine learning—ML, artificial intelligence—AI tools, and advanced algorithms). In this way, the IoT potential increases and introduces other development possibilities for BACS and BMS. Since the IoT edge nodes can increase computation near the source of the data, various IoT and cloud services can be deployed on local systems. This paradigm is known as 'edge computing' and integrates IoT technologies and cloud computing systems [45,46]. What is very important in smart home and building applications is that it reduces the communication bandwidth needed between sensors, actuators, and the external data center. Moreover, it allows for easier integration of different subsystems (energy, climate control, security, comfort, user services, maintenance, and energy management) controlled and monitored in modern, fully integrated intelligent facilities [3,39,47]. Therefore, this is one of the most important elements that should be included in the concept of a generic IoT framework for smart-building solutions.

A natural consequence of including edge modules from the IoT in BACS network structures, along with the computing and memory resources available in them (edge computing), is the emergence of a larger data exchange and processing structure called fog computing [45,46]. In [5], Taghizad-Tavana et al. explain that fog computing aims to optimize data transfer and communication between smart-building zones and smart homes and to develop lightweight algorithms to process local data and reduce the number of transmissions that are needed between devices. Moreover, according to [48,49] the fog computing paradigm is an alternative to smart modules with limited computational resources, typical for smart home and building systems. The authors explain that it extends the computational resources available in the cloud services to the network edge level, providing mobility, scalability, low latency, and robustness for the end BACS and BMS users. Additionally, what is very important is that edge computing enables real-time information analyses through the distribution of the decision-making process directly in the edge-level network at facilities (buildings, homes, local microgrids, etc.) [45]. Finally, Nasir et al. [39] add and explain that fog computing principally extends the cloud-computing architecture to the edge-level network. This approach enables an innovative variety of silent services and applications for end-users. Lightweight algorithms running on the edge-level network directly on IoT devices can conserve less bandwidth and provide computed, analyzed data to the end user without using the cloud every time. Moreover, the edge/fog modules can be equipped with AI mechanisms, providing more advanced computing and analyzing data in real time, thereby reducing the cloud service need and bandwidth. This approach and its features are very important considering the perspective of the development of advanced, dynamic control and monitoring functions for tactile internet, transactive energy management, generic IoT–fog–cloud BACS and BMS architectures as well as smart communities and cities [5,6,50,51].

1.3. Methodology of the Review

The wealth of conceptual and technical issues associated with the development of modern distributed smart home and building automation systems prompted the author to conduct a comprehensive review of the scientific literature of the last dozen years, particularly on the topic of integrating ICT networks and TCP/IP protocol transmission channels into these systems. The review is based on publication databases recognized in the electrical engineering, electronics, information technology and network control systems industry, namely ScienceDirect, Springer, IEEE Xplore, MDPI, and, occasionally, ACM Digital Library, Taylor and Francis, and Wiley Online Library. Moreover, in the selection of the main topics of the review, the results of analyses of the citations of publications in the Web of Science and Scopus databases and the population of selected keywords in patent databases (Google Patents search engine) were used.

Keywords were an important element in the guide to the literature review and in selecting specific publications for further discussion in this paper. They were divided into several thematic groups: (i) distributed control systems (e.g., building automation, fieldbus, smart home, smart building, BACS, and BMS); (ii) Internet of Things (e.g., smart nodes, edge computing, fog computing, cloud computing, TCP/IP for smart applications, and IoT maturity); (iii) communication networks (e.g., wired and wireless communication, remote access, local communication, and communication protocols); (iv) cybersecurity (e.g., privacy, data security, blockchain, access control, and encryption), and emerging trends (e.g., machine learning, artificial intelligence, tactile internet, and digital twin).

In many cases, combinations of several keywords led to the finding of other review texts, particularly those relating to IoT technologies and the areas of smart home and building applications. By analyzing these texts, the author identified potential thematic gaps and, based on them, formulated original contributions for this publication.

1.4. Original Contributions and the Paper Structure

According to the information presented in the previous subsections, IoT technologies have a very wide scope of applications. This review focuses on the specific smart home and building systems industries, considering the functionalities of advanced BACS and BMS. Moreover, several technical aspects of interactive energy management with DSM and DSR functions are discussed for smart home and building operations within local microgrids and smart-city infrastructures. In particular, the literature, research results, and case studies are analyzed in the context of developing the generic IoT concept for building automation systems in the framework of fieldbus–edge–fog–cloud architecture. The main contributions of this review are as follows.

- It provides a comprehensive review of the state-of-the-art IoT techniques and solutions related to smart homes and buildings with distributed control systems. This review is important because it collects knowledge about adapting and using IoT technologies in a segment that is rapidly developing but has so far been based on its own solutions for communication and data processing, in particular at the field level;
- As opposed to other IoT technology reviews, this one analyzes and discusses the suitability of various IoT concepts and tools for smart homes and buildings. Moreover, it sheds light on trends and innovative solutions emerging from this field that could be motivating for interested researchers and engineers;
- Providing a new perspective on various IoT applications (e.g., edge and fog computing and big-data processing) supported by recent research studies. To this end, the review provides some of the IoT design practices, considering the unique properties of smart homes and buildings, that finally will lead to more effective data processing, control and monitoring functions execution and better integration;
- It presents the major challenges and trends and pinpoints new, open research issues that need attention from researchers, domain experts, and engineers. In particular, this review provides information on the future scope of research on the integration of AI and ML capabilities, tactile internet developments, and IoT technology maturity assessment in building applications;
- It proposes general assumptions for the generic IoT framework concept with SWOT analysis as well as a discussion of pros and cons.

The remaining sections of this article are organized as follows. Section 2 presents a general view of network solutions, in particular distributed networks, used in home and building automation. Then, Section 3 describes the technological issues and main challenges related to the implementation of emerging edge and fog computing in BACS, BMS, and smart home and building systems. Section 4 selects and discusses several important trends and concepts for the development of functionally advanced BACS and BMS platforms with IoT technologies, in particular aspects of the implementation of ML and AI techniques. Section 5 introduces the generic IoT framework proposed by the author for applications in home and building automation systems, along with a SWOT analysis of its usability. Finally, in Section 6, the paper is concluded, providing future work information as well.

2. Control Networks and Smart Technologies in Buildings

In the classic engineering approach, technical solutions of smart home and building systems are based essentially on two organizational structures, (i) centralized systems, with one programmable logic controller (PLC) or server unit, cooperating with external modules supporting sensors and/or actuator modules and (ii) distributed systems, without a central unit but with sensor and actuator modules equipped with microcontrollers and communication interfaces. The second approach enables the execution of control and monitoring functions directly next to the elements of the house or building infrastructure, as well as data transfer between such modules, to implement more advanced functions within an integrated structure [52,53]. It should be noted that distributed solutions are the result of technological progress, development, and miniaturization of electronics, and they represent a significant achievement over the last several dozen years, enabling the implementation of more universal, advanced, and reliable system structures in the industrial and building automation industry. Obviously, centralized systems are still available and implemented in practice, but usually in very small installations (e.g., control of heating, ventilation, and air-conditioning systems in houses and small buildings), where they work well and are attractively priced. However, distributed systems are also becoming more and more popular in this sector. They are usually based on simple SoC modules with radio communication (Wi-Fi, Bluetooth, Z-Wave, ZigBee, and others), with dedicated applications for mobile devices or with support from dedicated server applications and websites [2,23,26,54,55]. Therefore, this study focuses on the analysis of the development of distributed automation networks, the architecture of which naturally fits into the concept of IoT technology applications and the cloud, with distributed tools and services for data processing.

2.1. Distributed Control Approach

The idea of distributed control systems in industrial and building automation developed in the 1990s and was a direct result of the appearance of microcontrollers that had sufficient computing power and memory resources to implement algorithms for the control and monitoring functions of industrial and building infrastructure devices. First, modules of various types of sensors and actuators were developed and equipped with microcontrollers and communication interfaces, necessary to exchange data (network variables and data points) between such modules. In the next stage of the development process, with the increase in the computing power and operating speed of microcontrollers (in the 2000s), universal programmable input–output modules appeared. Then, automation servers and other system modules have been introduced that support the processing of growing amounts of data at the object level [56–58]. As a result, especially in larger commercial and public buildings (e.g., hotels, shopping centers, offices, and university campuses), it became possible to build extensive automation and building management systems (BACS and BMS) with a fully distributed architecture. However, at the same time, the growing number of modules creating BACS and BMS forced the systematization of this architecture as well as the communication protocols. In [55,58,59], the authors describe and explain their most important elements, pointing to the progressive hierarchization of the BACS network architecture. According to the main assumptions, the overall architecture for a typical BACS can be organized into three layers/levels depending upon the functional hierarchy of the specific application:

- Field layer, the lowest one, where the interaction with field devices (sensors and actuators) happens, environmental data are collected, and parameters of the environment are physically controlled in response to commands from the system. Additional modules of sensors, stationary and mobile beacons with BLE, Z-Wave, and ZigBee wireless-communication technologies, are often used on this layer as well. They support mechanisms for the precise location of people and equipment in rooms and for monitoring environmental parameters;
- Automation layer, the middle layer, where data are processed, control loops are executed, and alarms are activated. It is also where processing entities also communicate values of more global interest to each other, and values for vertical access by the next management level are prepared (possibly aggregated);
- Management layer, the top layer, is where information from throughout the entire system is accessible, as well as where activities like system-data presentation, forwarding, trending, logging, and archiving take place. Moreover, vertical access to all BACS values is provided, including the modification of parameters such as schedules and long-term historical data storage. The possibility to generate reports and statistics is implemented as well.

With this concept, the progressive growth of data-processing tasks and services carried out at the highest Management Layer is clearly visible. However, the next stage of development of distributed BACS systems is not associated with progress in electronics, but with the rapid expansion of ICT networks, which over the last dozen years have been gradually reaching the management level and even lower levels in the architecture of BACS networks. Therefore, the key question is: how far this inclusion should go, whether TCP/IP protocols of ICT networks should dominate or perhaps completely take over communication and data handling in field-level networks? Are the protocols dedicated to field-level networks (ISO/OSI model), developed over many years and still strongly present in industrial and building automation and to be replaced by the expansion of TCP/IP, or should they exist together in a kind of symbiosis [20,60]?

In response, a broad, multiaspect analysis of the currently developed generic IoT concept is necessary, considering specific application requirements, security, privacy, and operational reliability of smart home and smart building systems.

2.2. IoT Structures and Technologies for Building Automation

The possibility of including TCP/IP communication channels in field-level networks has generated a multitude of application concepts at various levels of the existing architecture of BACS and BMS systems that address this topic by several engineering and scientific teams, along with additional marketing chaos. For example, the KNX Association and LonMark International, recognized organizations in the building automation industry, responsible for international open building automation standards, have launched information and advertising campaigns that the KNX [8] and LonWorks [9] standards are "IoT ready". Similar information still appears in the materials of many manufacturers of modules for BACS, BMS, and smart-home systems [61–63]. Therefore, in the first years of this process, research, engineering, and methodological works were already carried out aimed at verifying the technical capabilities of TCP/IP transmission channels and the IoT paradigm based on them in the effective implementation and support of the efficiency of BACS, BMS, and smart-home systems.

Scientific studies from the first two decades of the 2000s indicated potential areas of IoT applications and attempted to define possible development concepts for distributed networks. Kortuem et al. [64] propose a concept of smart objects as independent nodes with awareness (ability to sense, interpret, and react to events occurring in the physical world) and interaction (ability to converse with devices, other nodes, and the user in terms of input, output, control, and feedback). They discuss a general approach to such a concept, without specification of application areas (e.g., smart home or building), analyzing the possibilities of building peer-to-peer (P2P) data-exchange networks based on such smart objects to implement more advanced control and monitoring functions as well as data acquisition from sensors, actuators, etc. It is very important to note the reaction of smart objects to events, which is a key element of smart home and building systems, defined as event-based systems. In turn, in [65,66], the authors already point to the potential possibilities of IoT integration in the structures of BACS and BMS systems. However, the proposed applications concern only the use of IoT gateway modules and integration servers to support the operation of distributed BACS network nodes (integrated within field-level networks) in the implementation of remote access, data acquisition, and visualization in external services and object-linking and embedding for process control (OPC) databases. Therefore, IoT technologies in this approach constitute an addition, without significant interference in the structures of existing and planned BACS and BMS field-level networks. For example, the IoT with a TCP/IP protocol is considered a crucial element of a standardization process of building automation protocols.

However, the second decade of the 2000s brings more and more analyses and technical developments of the BACS architecture concept using IoT technology in system integration and development trends for modern smart home and building systems. In conference proceedings [67], Jung et al. discuss the new version of the IPv6 protocol and its most important features, such as the larger address space, self configuration, quality of service mechanisms, and security, qualifying it for applications in BACS and BMS, and promising a better integration of building automation technology in the IoT. Moreover, they conclude that the transition towards IPv6 from IPv4 at the *Automation* and *Management Layers* opens new opportunities for several previously not realizable use case scenarios in BACS and BMS like (i) home and/or building infrastructure device maintenance, (ii) smart grids and energy efficiency with interconnected devices, renewable energy sources (RES) and dynamic load shifting, the energy pricing ready for transactive energy concept, and, finally, (iii) buildings integrated into business processes with advanced occupants monitoring, access control, and HAVC operation and lighting control. In this context, several technical aspects of IoT integration with different BACS standards (KNX, LonWorks, BACnet, and OPC) are shortly discussed with a use case study.

Going one step further, Lilis et al. [68] proposed a transitional design for BACS networks that integrate IoT technologies. Based on the BACS architecture with field-level modules with communication in open standards (BACnet, KNX, and LonWorks), the use of the Internet backbone, and the developments in the embedded electronics at a higher level of the network structure, the authors point out the possibility and necessity of successive implementation of the embedded web services, sometimes referred to as Web of Things (WoT). In this way, the control and monitoring functions of BACS systems become services implemented in the form of applications in IoT devices at the automation or management layer, with communication of signals from and to field-level modules. Moreover, the authors analyze the practical possibilities of implementing openBMS-class platforms by providing a palette of semantic web services with the wide adoption of IoT-based management systems. For this purpose, they propose the implementation of universal distributed embedded electronics modules at the automation layer. According to this concept, each of those modules is an always-listening participant of the sensor and actuator networks and provides gateway-like capabilities towards the computer network [69,70].

Bearing in mind all development aspects of the concept of integrating IoT technology in BACS and BMS networks, Figure 1 presents their most important elements and differences visible, especially in the middle layer—the automation layer.

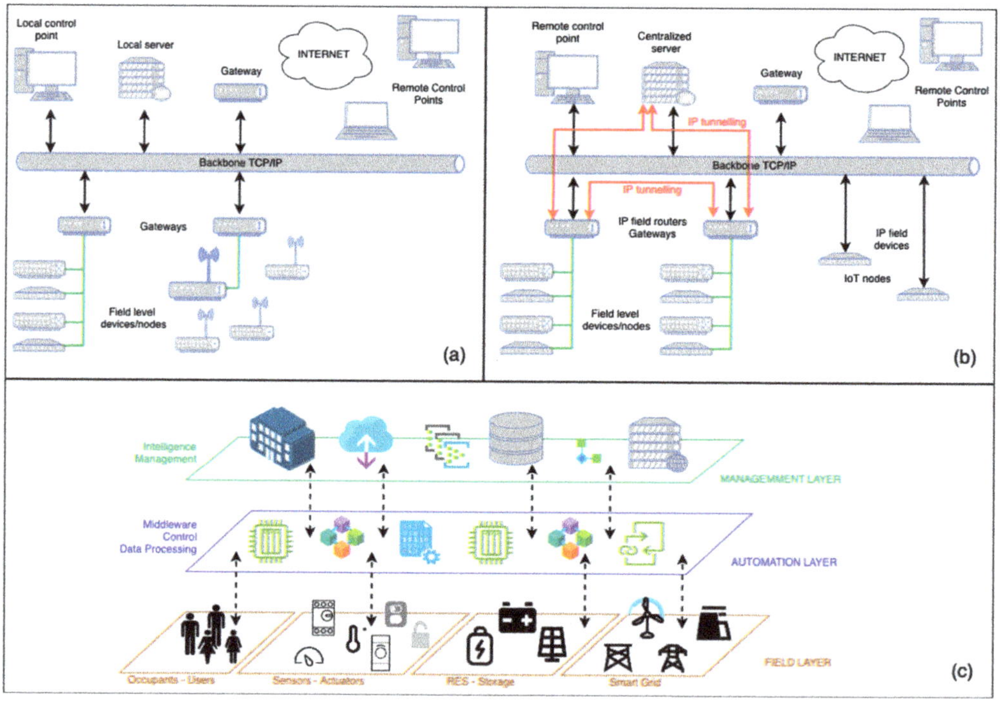

Figure 1. Schematic diagrams for two basic concepts of IoT development and application within the BACS and BMS architectures: (**a**) field-level devices connected by universal gateways to the higher network level using TCP/IP protocol and providing remote access [66]; (**b**) field-level devices connected by dedicated modules (automation servers, IP gateways) providing both remote access and data tunneling (IP channel integrated within the BACS and BMS architecture) [56,67]; and (**c**) additional distributed embedded electronics modules implemented at the automation level [68,69].

3. The IoT with Edge and Fog Computing in Buildings—Main Challenges

The turn of the second and third decades of the 2000s and until now is a period of rapid development of ICT network and cloud services. During this period, the widespread use of server resources (cloud) for storing and processing large amounts of data has been developing, basically, in all areas of industry, science, and social life. In the building automation industry, there are subsequent years of progressive integration of field-level networks with ICT networks and a trend toward implementing advanced control, monitoring, and management functions of an increasing number of elements of the building infrastructure. Moreover, the progressive implementation of energy-management algorithms, energy media, and the operation of local microgrids with RES and smart-grid services.

3.1. Service-Orientated IoT and Edge and Fog Computing in BACS and BMS

Such a significant development of functional concepts indicates new development trends in BACS systems in smart home and building applications. Simultaneously, the continuous development of IoT techniques and microcontrollers determines the need for organizational changes in BACS networks. In particular, this concerns the expansion of the ability to perform most of the analyzing and data-processing functions for monitoring and controlling the building infrastructure directly in the local network (within the building, campus of buildings, etc.). This is made possible by the computing power and memory resources of many modern distributed embedded electronics modules (automation servers, advanced routers, and gateways), integrated in the automation level of the IoT network. These modules, usually located at the junction of the field and automation (middle) layers, create the so-called edge computing in the modern BACS with IoT-network nomenclature [46]. Edge computing can be defined as a computing approach that uses resources in the periphery of a network. In this way, it brings the computation closer to the nodes of the BACS at the network's edge to provide a minimal delay and lower latency period between the moments that the data are acquired by sensors and then sent as control signals for actuators within the BACS [40,71,72]. The ongoing development of this layer of the BACS network, in particular the exchange of data in the TCP/IP channels between distributed embedded electronics modules and their performance of local, advanced analytical and data-processing functions at the automation level, has led to the creation of a new paradigm and term, fog computing, in the modern BACS with IoT networks. Fog computing is a distributed network resource that performs functions using local network resources but is also open to external services outside the local network—in the cloud [45,46]. Fog computing, therefore, operates at the automation and management levels, which are still supported by the local network and external resources. Hence, the fog element in the name indicates a kind of blurring of the integrated network layers [48,49,51,71]. The technical and organizational aspects of IoT networks presented in this subsection have significantly influenced the architecture of modern BACS and BMS systems in applications for smart homes and buildings. The general structure of such a network, highlighting the most important elements and levels, is shown in Figure 2.

In addition, BACS and BMS networks with such a structure, using distributed modules with TCP/IP communication in the edge and fog computing structure, create an environment for a service-oriented IoT [47] and a new Building as a Service (BaaS) [73] strategy. The first one is more general in nature; in the literature, it is referred also as Fog of Everything (FoE) and Internet of Everything (IoE) and focuses on the ability to use IoT technology in the implementation of services for four main areas: processes, data, people, and things. Formally, this approach refers to an ecosystem of edge modules that autonomously share and self-manage their limited resources in order to achieve the system goal (e.g., implementation of dynamic control, monitoring, management functions, etc.) [74].

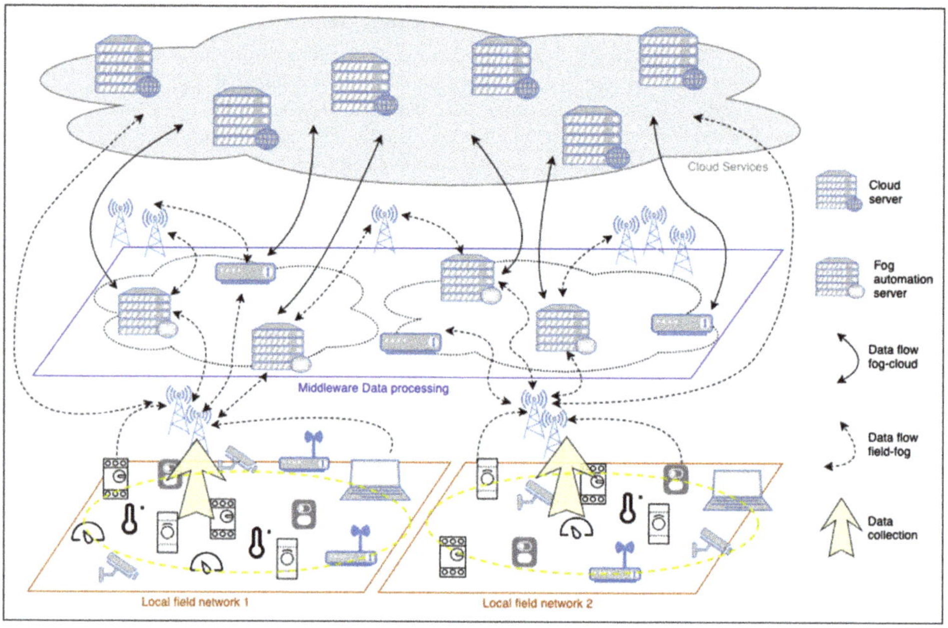

Figure 2. Structure and data flow in the network with field-level local networks, fog level, and cloud level [51].

The second one is more detailed and refers directly to the development concepts of BACS and BMS systems, in particular in smart-building applications. According to [75,76], buildings, in particular nonresidential ones equipped with BACS and IoT distributed networks integrated with fog and cloud computing, can be perceived in the BaaS convention, defined as the demand-oriented deployment of resources and assets, respectively. With this approach, buildings become platforms of information for providers and consumers. The focus moves from functions and services available in a building with BACS and BMS to view the building as a service-dominant logic-based asset. In this way, facility management (FM) is, in practice, a process of dynamic data management and data mining in order to adjust the operating conditions of building infrastructure devices to the current needs of users and changing environmental parameters (e.g., temperature, daylight level, energy tariffs, etc.). Moreover, it opens the way to building a framework of open data-processing platforms to provide specific services to users and infrastructure elements based on measurement data and device operating parameters.

Wildenauer et al. [75] also point to the inclusion of the BaaS and IoE approaches for enabling a digital twin (DT) tool based on building information modeling (BIM), which is becoming mandatory in several European states. In this context, it should be emphasized that the latest Energy Performance of Buildings Directive (EPBD 2018) [77] and the related technical report [78] define the smart readiness indicator (SRI) along with guidelines for verifying this readiness based on the services offered and possible to implement in the building. The first verification analyses of the usability of this indicator and related services in buildings are carried out as part of research and engineering works in order to develop mechanisms for applying the indicator's guidelines in real applications of buildings as well as energy microgrids with RES and energy storages [44,79–81].

3.2. Big-Data Processing and Cloud Computing

An aforementioned approach to BACS with IoT as a framework of an open data-processing platform requires the integration of numerous sensor and actuator modules as well as automation servers at the field and automation layers. Moreover, it is necessary to organize network connections of edge modules and computing infrastructure with external resources in the cloud. This entails the need to ensure efficient transmission and processing of large data resources while maintaining the time regime (real time) so that the implementation of BACS and BMS functions and services takes place essentially unnoticed by the building users. At the same time, in recent years, there has been a rapid increase in the popularity of data collection and processing services in the cloud—external servers usually operated by external entities or at the disposal of suppliers of smart home and smart building systems. This situation also affects designers and integrators of BACS systems with IoT, who often decide to implement cloud-centric systems, where there are basically only two levels of network structure, field and management (cloud) layers, and all more advanced functions and services in system are implemented in external cloud resources [82,83]. At the same time, they rely largely on data processing and protection tools offered by external administrators of such cloud services. However, this is not always beneficial, especially considering that many advanced services can be provided by modern BACS and IoT modules directly at the automation layer, close to the field layer modules. This solution naturally increases data security and reduces the load on network-communication channels. Therefore, in concept research and application case studies of modern BACS and BMS with IoT, solutions based on more advanced multilevel structures of system networks are considered and developed. The key element of these analyses is the development of guidelines regarding the areas of implementation of BACS and BMS functions and services in the network structure (what levels, between levels) and the methodology for the effective organization of network variables and data objects binding (interoperability, integration) to provide control and monitoring services. Considering the possibility of moving away from a cloud-centric organizational strategy, Chen et al. [84] propose an original cloud–fog computing architecture for information-centric IoT applications providing a classification of IoT applications and scheduling computing resources. Moreover, a developed scheduling mechanism optimizes the dispatch of cloud and fog resources regarding minimum cost in a cloud–fog computing environment. In turn, Sahil and Sood [85] discuss cloud–fog architecture implemented in a specific application, the panic-oriented disaster evacuation system in smart cities, with a particular analysis of the effectiveness of the proposed system-data-processing algorithms for various functional priorities (e.g., accuracy and sensitivity) in a very demanding time regime.

Research and development work is also carried out from a second perspective focused on the lowest levels of the network structure. In paper [4], the authors proposed a model and algorithms for handling modules with video cameras distributed at the field layer, with identification and classification services of recognized objects implemented at the automation layer in edge modules and a local workstation with the Microsoft Azure IoT Platform. Research focused on the functional capabilities of this solution and measurements of the system's effectiveness were carried out with the results discussion. In other studies, Huang et al. [47] propose an edge intelligence framework to build smart IoT applications. The project they developed is based on an extensive automation layer, with many edge modules cooperating to support local groups of field devices. A characteristic element of the concept is virtualized IoT services, which enable hardware-independent application design and simplify IoT services composition using different field-layer (physical) devices without redefining applications. This is an element of the ongoing strategy of organizing fog computing at the automation layer, within the local system network. Further development of the concept is proposed by Nasir et al. [39] by employing edge devices as a computational platform in terms of reducing energy costs and providing security, as well as remote control of all field devices and appliances behind a secure gateway. Moreover, in the automation layer, in addition to edge modules (nodes), they define fog nodes based on the powerful Jetson Nano device [86]. The platform is open for integration with external cloud services but is considered only as an additional tool to perform the most advanced processing, data analysis, and machine-learning services.

In turn, in paper [40], Lacatusu et al. analyze several design variants of the monitoring and control system for the infrastructure of a smart-buildings complex, based on edge computing and containers with additional cloud-computing services. Importantly, the authors conducted a comprehensive performance evaluation of design concepts using testing environments with two architectural options, (i) centralized (a cluster hosted in a public cloud) and (ii) decentralized (a similar cluster deployed in a local datacenter). They executed tests considering different numbers of edge nodes, corresponding to real application cases, namely a small apartment, a house, a small residential building, an office building, and a complex of smart buildings.

Finally, the research and engineering work of the last few years has focused on the development of various comprehensive concepts for the organization of smart home and building systems with the IoT–edge–fog–cloud architecture. For instance, in [3,51,87], the authors propose similar structures and frameworks for BACS and BMS networks with IoT, using in particular the new capabilities of edge and fog computing modules. In all cases, regardless of the application area, the structures of the automation layer are expanded, where operations are carried out by providing services such as data aggregation and analytics, security, access control, self healing, and self managing. The general diagram of such a network layer structure is shown in Figure 3.

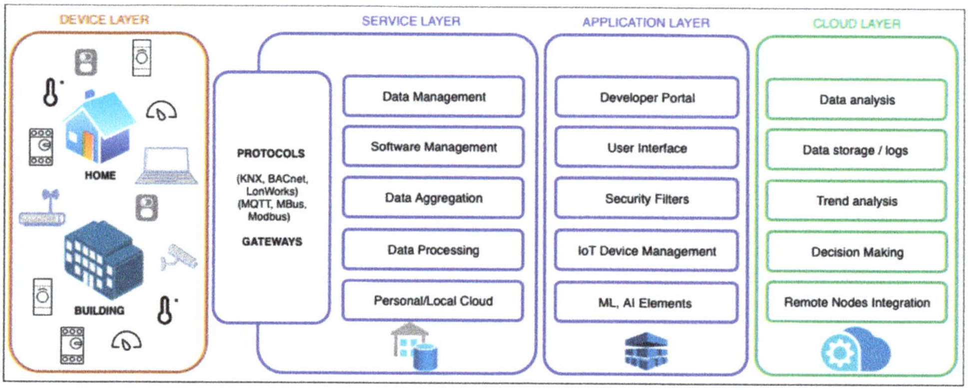

Figure 3. Advanced layer structure of BACS, BMS with IoT network, including big-data processing and cloud services [3,51,87].

For these solutions, the use of various communication technologies and the possibility of building network nodes based on universal modules with microcontrollers (e.g., Arduino and ESP) or a class of microcomputers (e.g., Raspberry Pi and Beagle Bone) are analyzed. Using the results of these analyses, engineering teams carry out tests aimed primarily at improving efficiency and reliability while rationalizing costs and resources used.

With this approach and the clear development trends of edge and fog computing in BACS and BMS systems, the issues of selecting communication protocol techniques and implementing data-security mechanisms, certainty, and unambiguity of communication become very important. In the context of the variety of available communication protocols, both wired and wireless, a comprehensive analysis of their usefulness and application potential was carried out in [71]. Furthermore, a broader analysis of security issues and data-transmission reliability was carried out in BACS and IoT edge computing networks in smart-city applications in [7].

3.3. *Cybersecurity, Privacy, and Blockchain Solutions for Distributed IoT in Buildings*

It should be noted that the aforementioned developments of new structural concepts of BACS and BMS networks in smart home and building applications, in particular the progressive distribution of IoT nodes and edge modules at the automation layer cooperating with external cloud services, resulted in a greater "openness" of the BACS network structure for new threats related to their inclusion and progressive integration into commonly used TCP/IP networks. Moreover, new structures of communication and access to data in the fog computing networks have been created, generating completely new categories of threats. According to [88], traditional conventional security mechanisms will not be designed or developed to secure technology such as the IoT. Therefore, it is necessary to develop and introduce innovative solutions in the field of data security and reliable trusted communication in the organized structures of a smart home and building network. These issues are the subjects of numerous research and technical analyses.

One of the most generalized analyses is presented in [38], where the authors indicate the most important issues related to the security and privacy of IoT networks. They discuss (i) confidentiality (data secrecy which guarantees the reliable transfer of data); (ii) data integrity (prevents corruption or alteration of data during transmission); (iii) availability/disposeability (ability to provide sufficient network and data processing resources when necessary), and (iv) authenticity (unique identification of users and resources authorized to operate on a given network). Moreover, they indicate significant challenges resulting from the development of IoT networks affecting security and safety issues. According to the authors, there are five main ones [38].

- Heterogeneity of devices and communication, resulting from the coexistence of various modules/nodes in one network structure (from small sensors and relays to large modules of automation servers and data servers), and the fact that they are produced by various manufacturers, often with different hardware architectures, supporting various types of software tools;
- Integration of physical devices; the result of the aforementioned 'openness' is that an attacker is potentially able to communicate with more devices than before. If he breaks the home/building/local network protection, he is able to manipulate the lighting system, lock doors, control HVAC, etc.;
- Constrained devices, the feature of many IoT devices resulting from a tendency to reduce the cost of their production. As a consequence, IoT devices have limited resources, memory space, low bandwidth, etc., and these considerably reduce the possibility to implement conventional security techniques;
- Large scale, since, currently, there are more computers and other IoT devices connected to the Internet than the number of humans on the globe, and the management of so large number of smart devices is a very demanding task and inevitably raises security risks;
- Privacy, IoT devices by their nature operate in a distributed structure, allowing communication for various wired and wireless technologies. This approach allows for interaction everywhere and data communication with many other BACS network nodes and edge modules in order to provide various services with different scopes and resource uses. The openness and flexibility of this structure generate additional privacy risks.

This is, of course, a very general summary. More threads emerge in detailed analyses. Particularly noteworthy is paper [89], where Parikh et al. consider security and privacy risks for all three of the most important levels of IoT networks, namely cloud computing, fog computing, and edge computing. The result of the analyses is a classification of the complexity of problems and preliminary proposals for solutions but without any technical or technological indications. In turn, paper [45] contains an overview of proposed solutions that increase the level of security and privacy in edge and fog computing structures. Laroui et al. provide a synthetic summary of the literature devoted to efforts to improve security and privacy in IoT networks, along with a brief discussion of the proposed models, mechanisms, and tools. Moreover, they discuss future research directions in this area considering the balance between openness and ease of use of the IoT networks and the need for a high level of their security and reliability.

From the point of view of BACS and BMS systems with the IoT, the most important are countermeasures dedicated to fog and edge computing integrated at the automation layer, usually within a local subnet. Such countermeasures are described by a detailed literature review by Alwakeel A. in [90]; in particular:

- For fog computing
  a. Encryption techniques;
  b. Decoy technique for authentication of data;
  c. Intrusion detection system for denial-of-service attack (DoS attack) [91] as well as port scanning attacks;
  d. Authentication schemes, where the fog computing network enables users to access the fog services from the fog infrastructure if they are well authenticated from the system;
  e. Blockchain strategy, it can prevent various malicious attacks in the fog network, including man-in-the-middle attack, DoS attack, and data tampering.
- For edge computing
  a. Edge node security;
  b. Full-time monitoring of edge nodes;
  c. Encryption with secret keys and attribute based [92];
  d. Intrusion detection system;
  e. User behavior profiling;
  f. Cryptographic techniques with smart secret keys;
  g. Data Confidentiality, for example with a privacy-preserving QueryGuard mechanism [93].

One of the most frequently discussed and analyzed solutions that is intended to support the implementation of the most advanced security and privacy elements is blockchain ledger technology [45,90]. In relation to the IoT paradigm, it is explained in [94] that blockchains, by definition, rely on a public directory acting as a common transaction information database for devices (nodes), edge modules, and automation servers. Furthermore, in [95], Moniruzzaman et al. discuss the blockchain-based smart-home ecosystem, with the framework presented in Figure 4. According to them, it is a four-layer conceptual framework consisting of four layers, namely the (i) IoT data sources layer, (ii) blockchain network layer, (iii) smart-home applications layer, and (iv) clients layer.

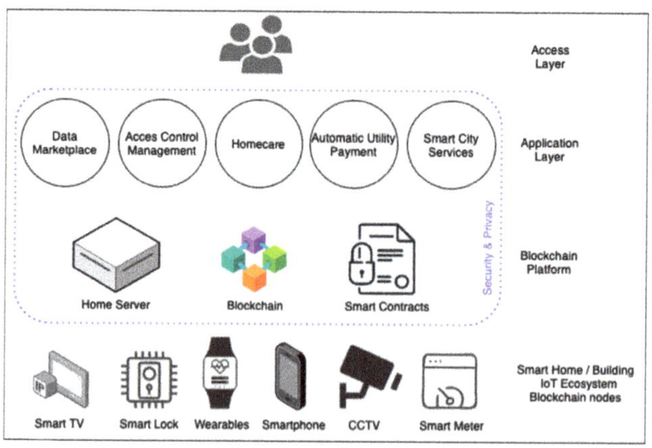

Figure 4. A four-layer application framework of a blockchain-based smart-home ecosystem proposed and discussed in [95].

Sensors and actuators located in the first one generate and/or use data consolidated and stored in edge modules (servers) or a decentralized platform such as the second one—a blockchain. All of the events and acts of the sensors and actuators became smart transactions used to realize services. What is characteristic is that time is an indestructible database that is placed in a new transaction and divided into a block hash chain. In this way, many copies of blocks are made and saved in the extracted node protocol. Moreover, hash values cryptographically connect blocks, and edge modules (servers) may be considered miners, which are responsible for verifying and adding new transactions to new blocks while smart contracts follow predefined rules and facilitate the decentralized transactions [94,96]. This organization of data processing as a transaction with a trace in the block structure fits naturally into the framework of distributed BACS and BMS with IoT networks [95]. Additionally, it opens the way to easier and more reliable integration with external services, for instance in the community microgrid frameworks suggested in [97].

Importantly, the more distributed the network nodes in such a structure, the higher the security level due to blocking verification procedures in the nodes. Therefore, the distribution factor, previously identified as reducing data security, becomes an advantage with this approach. The pros and cons related to the implementation of blockchain technology in IoT networks in various application areas, including smart homes and buildings, are discussed in [98,99], considering security and privacy aspects as well, and indicating the added value of such an approach. A detailed analysis of the transaction workflow along with the accompanying tools and methods of data protection in the fog and edge computing network structure, is presented in [100]. In the conclusion section, the authors also provide a comprehensive review of research work focused on the possibility of increasing the level of security and privacy in IoT networks, along with an indication of various limitations. Some of the latest research suggests innovations in the integration of blockchain technology in IoT networks, allowing for overcoming the limitations of the classic approach, such as scalability, storage and bandwidth, transaction charges (checking by miners), data privacy (sharing every node), and network size (all nodes within the network). In [88], Alshaikhli et al. introduce an IoT application (IOTA) distributed ledger technology that can provide unlimited scalability specifically suitable for the IoT with fog and edge computing. In particular, this technology provides fully distributed data transactions without a central authority unit, microtransactions in real time with zero fees, a new scalable distributed ledger mechanism, and masked authentication messaging with advanced encryption of data.

4. New Ideas, Concepts and Trends

The generic IoT concept for smart buildings and field-level automation should be considered first of all in the context of needs and facilitations in organizing and integrating increasingly distributed network nodes as well as new ideas and development trends of smart building systems with IoT. Since IoT technologies and application areas are still undergoing rapid development in many areas, this study selects several of the most important aspects that seem to be important in relation to smart home and building systems.

4.1. Machine Learning and Artificial Intelligence

The development of modern techniques for collecting and processing big data has allowed for the effective implementation and use of ML and AI mechanisms, in particular supporting the organization and functioning operation of automation systems. According to Djenouri et al. [101], in the context of BACS and BMS, ML techniques could be used to solve fundamental problems, such as predicting occupant behavior and preferences and forecasting energy demand and peak periods, which are difficult to solve with traditional programming, but potential solutions can be achieved from advanced and fast data analyses. They reviewed several research and studies and discussed potential areas for ML applications in two aforementioned categories:

- Occupant-centric solutions
    a. Occupancy detection, prediction, and estimation providing essential information for advanced control of several subsystems like HVAC;

b. Activity recognition to provide better control scenarios, tailored to increased or limited user activity, e.g., in different zones of the building;
c. User preferences and behavior to provide well-tailored thermal and lighting comfort, considering individual or group user preferences, as well as operating scenarios for home devices and building infrastructure tailored to the most common recurring user behaviors;
d. Authentication schemes, where a fog computing network enables users to access the fog services from the fog infrastructure if they are well authenticated in the system;
e. Blockchain strategy, it can prevent various malicious attacks in a fog network, including man-in-the-middle attack, DoS attack, and data tampering.

- Energy/device-centric solutions
    a. Energy profiling and demand estimation in the context of using BACS and BMS monitoring and control functions to improve the energy efficiency of buildings, in particular, those incorporated into the structures of local energy microgrids and prosumer installations [12,102];
    b. Appliance profiling and fault detection to track and identify different buildings' appliances, as well as detect anomalies/failures in the different components of the energy management system. Moreover, this approach allows support of the DSM and DSR mechanisms of transactive energy [103,104].

These two categories are mostly discussed in the context of ML applications within smart homes and buildings. In [105], the authors analyze in detail various technical and functional aspects of human activity recognition in smart homes using algorithms for IoT sensor networks, considering the pros and cons of different ML methods and tools dedicated for various smart home and building applications. However, in [106], Suman et al. point out that, in turn, advanced IoT and BACS devices may impact the behaviors of people in buildings. Based on human and various thermal and environmental models, the authors analyze their possible mutual impact, in particular, changes in human behavior depending on changes in building infrastructure control scenarios and comfort parameters.

In turn, in [107] Machorro-Cano et al. present a HEMS-IoT, a big data and machine learning-based smart-home energy-management system to provide home comfort, safety, and energy savings. ML techniques and big-data processing technologies are important in this solution since they help to analyze and classify energy consumption efficiency, identify user behavior patterns, and offer increased comfort at home with rational energy usage. Additionally, in [108], the authors identify the most essential BACS with IoT-enabled factors that sanction a need for ML, as well as AI integration with smart homes and buildings to provide energy-efficiency improvements and facilitate energy management. Research, analyses, and case studies are carried out in this area using advanced functionalities and communication techniques [41,80,109–111].

Another issue is the possibility of using ML mechanisms with AI elements to recognize, classify, and service BACS and IoT modules and network nodes. Cvitic et al. [112] propose an original approach and an ML-based IoT device classification model considering various sets of data and different data traffic models. Furthermore, considering the growing use of BACS with IoT solutions, especially with edge and fog computing, Huang et al. [47] note that real-time detection of unexpected, emerging, or spontaneous situations is important for increasing the reliability of the network and improving its maintenance. This approach makes recent data more valuable than historical data for the learning models, which also determines the need to develop ML mechanisms with a shorter time window for analyzing data sets. All these issues indicate the growing importance and even indispensability of ML technologies and methods in BACS, BMS, and IoT systems in the coming years. Moreover, research is already being conducted to develop new trends in ML development. Due to the increasing computing power of edge and fog-level network nodes, a federated learning (FL) approach is proposed [50,113,114]. In this concept, the nodes within the IoT network get involved in the training and inferring process, keeping the raw data within themselves and sending only the results of local training processes performed on these network nodes, to maintain privacy and reduce communication overhead. Importantly, FL mechanisms based on the dispersion of network nodes and their computing power are indicated as important elements of the development of blockchain technology in the field of more advanced data security and privacy mechanisms in BACS networks with IoT [115–117].

However, the AI functions and solutions are particularly considered in the context of support in the integration processes of extensive BACS system networks with IoT, supporting very diverse functional and infrastructure subsystems of buildings and homes. First of all, AI integration is important since, in classic BACS and IoT networks, the design and architecture development of each control function, and the rule only works in one subsystem (e.g., HVAC, lighting, security, etc.); there is no interoperability between these subsystems (or it is very limited). Furthermore, as previously stated in Sections 1.1 and 1.2, the monitoring and control functions of these systems are often aided by other modules such as sensors and beacons with wireless-communication interfaces that do not support the TCP/IP protocol directly. This requires the use of additional gateways or data concentrators.

Considering this, the model proposed in [3] facilitates and allows the integration of new digital services based on BACS and IoT nodes, providing deeper interoperability of the different subsystems and introducing new services based on ML and AI techniques to homes and buildings. The authors have implemented the model and verified it in tests. Moreover, in [118], Panchalingam et al. describe several smart-building domains that should be considered for integration with AI techniques in relation to those techniques. They suggest and discuss what research on AI techniques should be conducted to improve safety, BACS and IoT systems design, control logic, and energy efficiency in buildings as well. The similar aspects are analyzed in [119,120], considering not only functional and organizational aspects, but technical and architectural as well.

The synthetic summary in graphical form in Figure 5 indicates the areas in which the use of ML, FL, and AI techniques is observed and suggested.

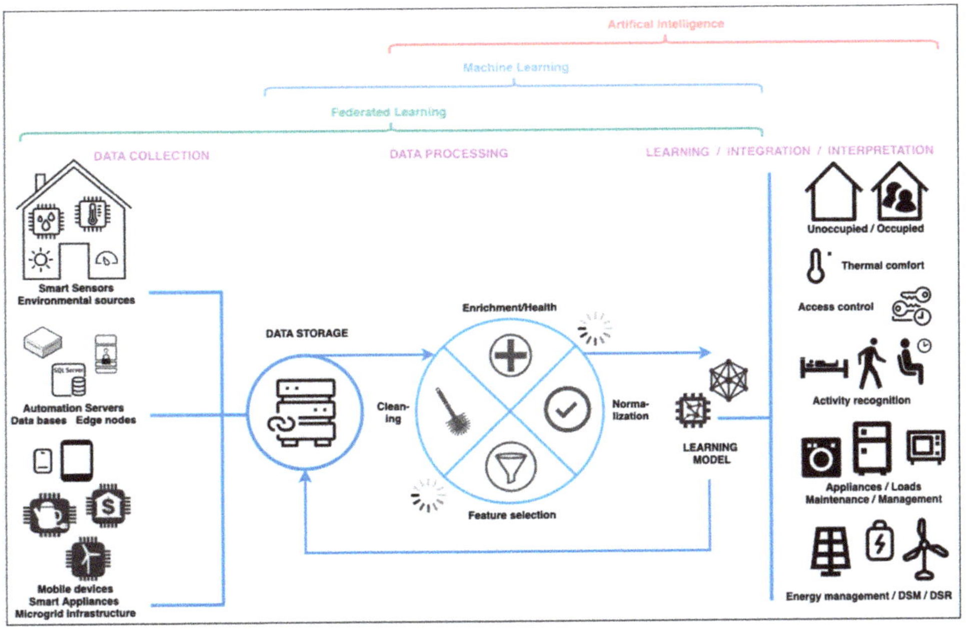

Figure 5. The BACS with IoT systems areas for implementation of ML, FL, and AI techniques, methods, and tools (based on [101]).

4.2. Tactile Internet and Digital Twins with Distributed Automation Networks

All these methods and technologies, namely ML, FL, and AI, become the basis for the implementation of new functional possibilities and the development of emerging trends of BACS and BMS systems with IoT. The author has selected two, namely Tactile Internet and Digital Twins, which are, in his opinion, currently the most important trends that are both part of the development of a new philosophy of using smart home and building systems and as operational maintenance support techniques, especially for large BMS with IoT systems in smart buildings. The importance of the emergence of these development trends is indicated by the increase in the number of scientific publications observed since the mid-second decade of the 2000s, in particular those resulting from research and development projects. In the publication databases of ScienceDirect (Elsevier), Springer, and IEEE Xplore, 80% of publications on the topic of "tactile internet in smart applications" are in the years 2017–2023. Of which, the years 2020–2023 amount to almost 300 publications per year. A similar proportion applies to the recognized bibliometric services Web of Science and Scopus. In turn, in relation to the second emerging trend, Digital Twins in smart buildings, the analysis of the number of scientific publications in the ScienceDirect, Springer, and IEEE Xplore databases indicates an even narrower time spectrum, 2019–2023, with a rapidly growing number of publications (for example in ScienceDirect in 2020: 462, in 2021: 714, but in 2022: 1094, in 2023: 1463). In the Web of Science and Scopus bibliometric services, the first single publications on this topic were recorded in 2017–2018, and in 2020–2023 there are already almost 200 publications per year.

The first of the discussed emerging trends is the Tactile Internet or Tactile Internet of Things (TIoT), considered the second generation of the IoT to support the transfer of haptic data (what is sensed by the skin) and kinesthetic data (muscle movement), in addition to audio, video, and images as tools for the human–smart home system interface [50,121]. In its most basic approach, TIoT involves wireless communication (5G and Wi-Fi) and classic wired channels to control real and virtual objects (actuators) by humans in real time. By enabling the control of the IoT nodes in real time, it also provides haptic sensations to create a new extent for human–machine interaction in homes, buildings, and industry [122,123]. The assumptions of the technical organization and architecture of TIoT systems are currently the subject of research and development work, but, as Fanibhare et al. [122] point out, crucial design goals can be achieved by placing TIoT nodes close to each other, which is possible with a distributed and decentralized architecture dependent on recent technological advancements, such as edge and fog computing. Therefore, modern, fully distributed, BACS and BMS installations with elements of fog and cloud computing services are becoming a natural implementation environment for TIoT. In particular, the development of user interfaces based on virtual reality (VR) and advanced applications for monitoring the activity, behavior, and health parameters of occupants is expected [122,124]. However, the implementation of user interfaces and functions in the haptic and immersive real-time interaction regime introduces new requirements for the data communication network, both in terms of its speed and throughput. In TIoT applications, a response time to events of 1 ms is required, much shorter than in the case of audio (100 ms) or video (10 ms) interfaces. That is why fog computing and FL technologies are becoming so important for the effective implementation of the TIoT concept, supporting mechanisms of local processing of larger data volumes and transmission of the results of analytical tools [50,121–123].

The second selected emerging trend is the Digital Twin (DT) environment, the concept of which is being developed for many industries and the building industry. It is usually discussed in relation to BIM techniques, which are based on technical data and operational parameters of the building for the purpose of modeling its architectural, installation, and utility structure. According to [125,126], BIM is used in architecture, construction, engineering, and facility management to facilitate the planning and analysis of various scenarios and building organization concepts as well as clash detection, lean construction, cost, and time estimation. However, the BIM concept does not include the element of data dynamics and the related predictive capabilities. In [125] the authors point out the two most important differences between DT and BIM, namely (i) the BIM was designed to improve the efficiency of design and construction, and still is used in these processes, but the DT is designed to monitor physical assets and improve their operational efficiency and provide predictive maintenance and (ii) the BIM was not designed to work with real-time data; therefore, it is used for design and construction facility management, whereas the DT is a dynamic environment, with support for real-time data and ML and AI. Moreover, in [127], Hadjidemetriou et al. describe building DT architecture separating four phases:

1. Collection of data and information regarding the geometry, materials, and equipment characteristics of the specific building of interest. This information is necessary for modeling the building;
2. Collection of live measurements from sensors and electrical meters installed in the building to monitor its real-time operating conditions. In this context, modules with wireless-communication protocols, such as BLE, Z-Wave, and ZigBee, can be used at the field level, along with the required infrastructure for integration into the IoT TCP/IP network [128–131]. Additionally, live weather data could also be collected. These live data are directly incorporated as inputs into simulation tools to replicate the building's operating conditions in real time;
3. Simulation tools with model-based modeling are incorporated to simulate building control and monitoring systems. Intelligent algorithms can also be used to calibrate the building parameters in order to achieve better comfort and/or improve energy efficiency;
4. Development of a software platform to integrate the three previous phases. That platform is responsible for the proper data exchange and the successful real-time execution of the simulation tools as well as for integrating monitoring and control applications and investigating different what-if scenarios.

This architecture is presented in the graphical form in Figure 6, published and described in detail in paper [132]. Moreover, the dependencies between the BIM and DT techniques, in terms of their use in distributed systems with IoT, are analyzed in paper [133].

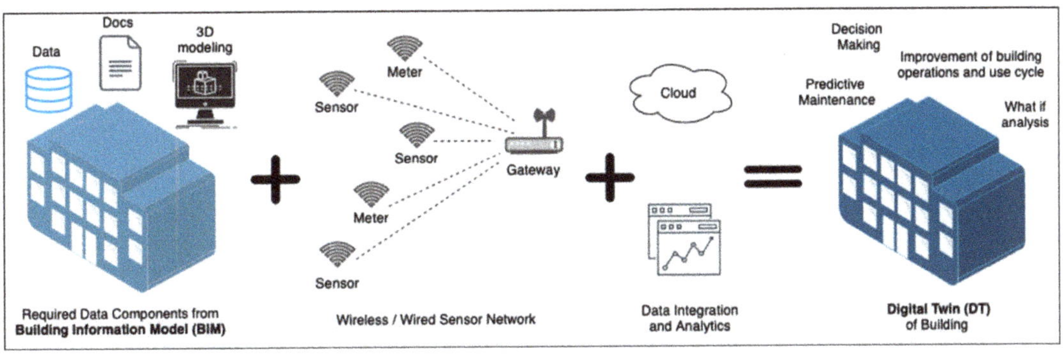

Figure 6. Essential components of building DT architecture [133].

From the analysis of the DT concept for buildings, it can be clearly stated that this is an environment requiring the involvement of all existing technologies and development tools of distributed BACS and BMS networks with edge, fog, and cloud computing elements. Thanks to them, it is possible to collect information about the operating status of the building's infrastructure, users' behavior and activity, and conduct active energy-management mechanisms, demand and load prediction, and provide control and monitoring functions [125,126].

4.3. IoT Technology and Maturity Assessment

The multitude of technical solutions and potential frameworks of modern network-automation systems with IoT technologies add complexity and raise questions concerning facility management strategies, organizational structures, and technological capabilities in the implementation of basic or advanced functions of monitoring, control, management of buildings, home infrastructure, etc. Additionally, many existing and operated buildings are equipped with very diverse IT systems, field-level networks, proprietary control systems, and other platforms supporting building management and BMS tools. In such a situation, the transition to modern IoT technologies and practices in order to streamline and improve the capabilities of effective building management is difficult and, above all, requires preliminary sorting and assessment of BACS, BMS, IoT technologies, and tools available to the user or building manager.

Therefore, in recent years, efforts have been made to develop IoT readiness assessment methods and tools. They focus, in particular, on the evaluation of two areas, namely the technical and organizational conditions of network systems in buildings, in terms of the possibility of their use in the development of infrastructure for comprehensive smart home and building systems with IoT. In [134], Arsenijevic et al. describe four possible methods for assessing the technological maturity of the IoT with varying levels of detail. The most important verification factors were analyzed, in particular, those related to the network structure (centralized or distributed), available computing power and data analytics tools, diversity of standards, and data-transmission protocols in the system, and also the readiness of the IT team to support new networks with edge and fog elements and cloud computing. In turn, in paper [135], Metwally et al. analyze these methods in detail, along with additional technical and organizational aspects relating directly to IoT applications in BACS and BMS. As a result, they proposed their own scale and indicator for assessing IoT readiness, with five levels of advancement:

1. Low IoT level, larger manual, low automatic control at the building level (local automation);
2. Mid-IoT level, automatic control at the building level (centralized automation), firstly emerging of DALI controls for lighting as well as field-level sensors for some control functions;
3. High IoT Level, automatic control at the building level (distributed automation), with networked sensors and modules and nodes to control most systems' functions with the performance analysis;
4. Fully IoT level, automatic control across all buildings/site levels (distributed networked automation) with networked sensors, all modules and nodes to control most systems' functions with the performance analysis also perform a predictive decision making.

The authors of paper [1] where, after a comprehensive analysis of existing methods to verify technological maturity and readiness for IoT solutions, proposed a four-level IoT assessment model, but with an additional level of zero. This model, however, relates mainly to organizational issues of preparation of staff and teams operating the network infrastructure and their awareness of system transformation, and to a lesser extent to technical and technological issues; although, of course, it does not ignore them.

In the context of the development of BACS and BMS platforms with IoT, it should be emphasized that the mentioned models and indicators complement the standards and studies regarding the selection of basic and advanced functions of home and building automation—standards EN 15232 and ISO EN 52120 [136,137]—and the assessment of the readiness of buildings for intelligent solutions and smart grid networks with the Smart Readiness Indicator (SRI)—EPBD directive [77]—and technical report [78].

5. Generic IoT Framework—Concept, Development, and Discussion

Bearing in mind all the technical and organizational aspects analyzed in Sections 2–4 of this review, the author proposes systematizing the most important elements relating to the technical, organizational, and conceptual issues, from the perspective of developing the concept and implementing the so-called generic IoT [138–140]. Wang W. et al. in [138,139] discuss for the first time the general concept of generic IoT, focusing on optimization issues and essentially reducing the size of data necessary for transmission between IoT network nodes. The approach they have developed and tested allows for more effective data handling by devices with limited resources and computing power. Moreover, they indicate that achieving integration both on a device and semantic (data) level for physical objects and services is possible thanks to the virtualization of middleware environment objects (edge and fog computing). With this approach, the handled data objects and integrated network nodes become universal, increasing the freedom of their connection and processing. A similar strategy for developing generic IoT is undertaken by Ali Z. et al. [140] who developed the thread of implementation of a number of data-processing services and information about network modules in the middleware environment (data acquisition, device heterogeneity, service management, security and privacy, interoperability, scalability, flexibility, data processing, and visualization). Considering the rapid technological progress and the increase in the possibilities of local data processing of edge and fog network nodes, they discuss for the first time the possibilities of implementing advanced data-processing mechanisms, including AI functions, in the middleware environment. They verify their proposals by analyzing the results of implementing the proposed mechanism in a smart-city application. However, all the publications discussed above indicate the dependence of the concept of generic IoT on many different factors, including policy, standardization, and development of innovative technologies (research and development), conditions, and requirements of specific applications, as well as the technological possibilities of supporting increasingly advanced mechanisms and algorithms for data handling at the middleware and object level.

Therefore, since designing an advanced framework for generic IoT systems in the context of building automation and smart-home systems involves careful consideration of various elements to ensure seamless integration and optimal functionality, the author of this paper decided to review and consider them, proposing holistic generic IoT framework dedicated for this type of applications. In the next subsections, there is a structured framework proposed, outlining both the mandatory and optional elements, as well as considering specific requirements for smart home and building applications, including edge, fog, and cloud computing. Moreover, the main fields of potential research and development work are suggested as well.

5.1. Mandatory Elements of the Framework

The elements collected in this group are crucial for the generic IoT framework due to their fundamental roles in ensuring the effectiveness, reliability, and security of the entire system with BACS and IoT nodes. They are divided into six levels.

Device layer: sensors and actuators form the foundation of the field level within the network, enabling data collection and control, which are essential for smart decision-making both in building automation and smart homes. All international BACS standards, other standardized protocols (e.g., message queuing telemetry transport—MQTT, constrained application protocol—CoAP), and wireless-communication technologies discussed in Sections 1–3 should be considered for implementation at this layer.

Communication layer: standardized protocols and gateways facilitate seamless communication between diverse devices, providing interoperability and efficient data exchange. They should be considered for all network layers discussed in Section 2, implementing again the MQTT, CoAP, and additional real-time technologies and protocols.

Data-processing layer: Edge computing enhances real-time data processing near the device level within a local network, reducing latency and ensuring timely responses, while data filtering and aggregation optimize network resources. Considering fog computing and even integration with cloud computing, it is recommended to conduct research and development work with Amazon Web Services (AWS) IoT Greengrass, Google Cloud IoT Edge, and Azure IoT Edge.

Integration layer: BIM and DT based on the TCP/IP protocol and middleware enable the harmonious integration of IoT devices with building structures and diverse systems, promoting a cohesive and interoperable environment. It is suggested to conduct research and development work considering RESTful Application programming interfaces (APIs), MQTT, CoAP, etc., tools and protocols to develop standardized APIs and middleware to enable communication and data exchange between different IoT devices and systems, ensuring interoperability. An exploration of new service discovery algorithms should be mentioned as well. Their development and implementation would support the dynamic discovery and registration of IoT services and resources, facilitating the integration of new devices without manual configuration.

Security layer: end-to-end encryption and access controls are paramount for safeguarding sensitive data, ensuring the integrity and confidentiality of information in the BACS and BMS IoT ecosystem. Considering the openness of the IoT networks, developments of this layer should be considered first of all Intrusion Detection Systems (IDS) and Intrusion Prevention Systems (IPS) aimed to identify and respond to potential security threats, enhancing the resilience of the IoT ecosystem. Moreover, implementation of end-to-end encryption with, for example, Transport Layer Security (TLS) and Datagram Transport Layer Security (DTLS), should be examined, to secure data communication between IoT devices and the cloud, preventing unauthorized access and ensuring data confidentiality.

Cloud-computing layer: leveraging cloud storage and ensuring scalability supports the archiving of historical data and large-scale analytics and accommodates the evolving nature of generic IoT systems. There are many cloud services and tools that could be developed for this type of application, for example, AWS Lambda, Microsoft Azure Virtual Machines, Google Cloud Firestore, etc., with the aim to host applications and services that require computing resources, facilitate data processing, and application deployment.

It should be noted that these mandatory elements collectively establish a solid foundation for a reliable, secure, and integrated generic IoT framework. They address the core aspects of device communication, data processing, integration, and security, providing solutions for the successful implementation of advanced features and technologies in smart home and building applications with integrated IoT technologies.

5.2. Optional Elements of the Framework

The elements collected in this group enhance the generic IoT framework by introducing advanced capabilities that address the specific requirements of smart home and building applications and the overall performance of the generic IoT network. Considering that, they are presented in two subgroups, related to the smart home and the smart building.

5.2.1. Smart-Home Applications

Remote access and control: the development of mobile applications to provide homeowners and users with remote access to monitor and control smart-home devices; the implementation and integration of voice commands for convenient hands-free control. This application area is important for low-energy wireless-communication technologies such as BLE, ZigBee, and Z-Wave.

User interface: dashboards and control panels enable intuitive interfaces for homeowners to monitor and control smart-home devices effortlessly; moreover, customization of the user interfaces allows them to personalize automation rules based on their preferences, enhancing the user experience. In this field, several technical concepts could be considered for research and development, like voice command integration, with, for example, Amazon Alexa Skills Kit, Google Actions, etc. Moreover, web and mobile app development frameworks should be utilized to build responsive and interactive user interfaces for smart-home applications accessible through web browsers and mobile devices. Augmented reality will probably also be a new and emerging trend in organizing modern user interfaces.

Energy efficiency: integration of energy monitoring devices to empower homeowners with insights into energy consumption, promoting energy-efficient practices; smart grid integration explores connections with smart grids for optimized energy management within the smart-home environment, using DSM and DSR functions and tools. The most important directions of technological and systemic development seem to be (i) smart energy meters and IoT-enabled power outlets, allowing for the integration of energy monitoring devices to track and analyze the energy consumption of individual devices and appliances within the smart home; (ii) integration with local smart grids in relation to RES and energy storage use to optimize energy consumption, leveraging real-time data to make informed decisions about energy usage; and, last but not least, (iii) integration with dynamic pricing platforms (transactive energy mechanisms) implementing systems that adjust energy consumption based on dynamic pricing models, allowing users to optimize energy usage during periods of lower electricity costs.

5.2.2. Smart-Building Applications

Fog computing: local data-processing nodes deploying *fog computing* for smart-building applications, supporting local data processing for reduced latency and enhanced responsiveness in large-scale systems. Considering the fog computing layer/element, various tools and solutions can be employed to enhance real-time processing capabilities at the edge of the network. However, research and development are primarily suggested in the areas like leverage edge computing platforms (e.g., AWS IoT Greengrass, Azure IoT Edge, and Google Cloud IoT Edge) to extend cloud capabilities to edge devices, enabling local computation, data storage, and execution of IoT applications as well as the development of lightweight algorithms optimized for edge computing that are resource-efficient and well-suited for edge devices to enable real-time processing without compromising performance. Containerization is also an important emerging element that appears in the analyzed concepts for the development of fog computing for smart buildings. It provides tools to package and deploy applications consistently across edge devices, facilitating efficient deployment and management of fog computing resources.

ML and AI: utilizing ML for predictive analytics in smart-building management, optimizing resource allocation and improving overall efficiency and implementing of AI-based anomaly detection for proactive identification of faults and irregularities in building automation systems. Currently, it is a very dynamically developing field. The suggested main directions of research and development of ML and AI applications in smart building applications are (i) predictive maintenance models with ML algorithms that predict when building equipment and systems require maintenance, minimizing downtime and reducing operational costs; (ii) energy-consumption forecasting employing AI models to forecast building energy consumption, enabling proactive energy management and cost optimization with DSM and DSR mechanisms; and (iii) exploring of reinforcement learning techniques for building automation, allowing BACS and BMS systems to adapt and learn optimal control strategies over time.

Regulatory Compliance: ensure robust data-privacy measures to comply with regulations, addressing the unique challenges associated with handling sensitive data in smart-building applications and compliance with energy-efficiency standards where specific energy-efficiency standards applicable to commercial and large-scale buildings must be complied with. This area depends largely on institutions and nontechnical conditions. But, first of all, new regulations are expected in the field of data privacy, with a focus on protecting the personal information collected by smart-building systems, and updates to cybersecurity standards for IoT and smart buildings to address evolving threats and vulnerabilities. Moreover, establishing interoperability standards for smart buildings, ensuring compatibility and seamless integration of diverse devices and systems should be considered as well. In this context, regulations and standards for new smart-city platforms and frameworks are expected to promote the cohesive development of smart homes, buildings, and microgrids and the deployment of IoT technologies in these applications.

It should be noted that all elements from both subgroups can be mixed, being used in both smart home and building applications. However, he points out that some of them are dedicated only to specific applications, for example, regulatory compliance is specific to larger buildings.

5.3. SWOT Analysis and Discussion—Main Challenges, Opportunities, Pros, and Cons

The usefulness of the presented generic IoT framework requires an analysis of the possibilities and challenges arising from its potential implementation and possible difficulties as well as threats in its practical implementation in a smart home and BACS and BMS with IoT installations. Therefore, the author decided to present the SWOT analysis, along with a short discussion.

S—Strengths:

- Comprehensive integration: the incorporation of mandatory elements from the framework ensures a solid foundation for seamless device communication, data processing, and security;
- Flexibility and scalability: the inclusion of optional elements allows for customization based on specific applications, catering to the unique needs of both smart homes and buildings;
- Advanced capabilities: optional elements such as fog computing, machine learning, and AI enhance the framework's capabilities, providing predictive analytics, anomaly detection, and efficient resource management.

W—Weaknesses:

- Complex implementation: the inclusion of various optional elements may introduce complexity in the implementation phase, requiring careful planning and expertise;
- Resource intensiveness: certain advanced features, such as ML and AI, may demand substantial computing resources, potentially affecting system performance;
- Potential security risks: the complexity of the framework may introduce vulnerabilities, necessitating robust cybersecurity measures to mitigate potential risks.

O—Opportunities:

- Market growth: the rising demand for smart home and building solutions, as well as IoT and TIoT, presents a significant market opportunity, with the framework well-positioned to capitalize on this trend;
- Technological advancements: ongoing advancements in IoT technologies, including edge, fog computing, ML and AI offer opportunities for continuous improvement and innovation within the framework;
- Regulatory support: compliance with emerging data-privacy and energy-efficiency regulations can enhance the credibility of the framework and market acceptance.

T—Threats:

- Cybersecurity concerns: as IoT systems become more interconnected, the framework faces potential threats from cyberattacks, necessitating robust security measures;
- Integration challenges: compatibility issues with existing systems in buildings or homes may pose challenges during implementation, requiring seamless integration strategies;
- Market, research, and technical competition: rapid technological advancements may lead to increased competition, requiring continuous updates to maintain the framework's competitiveness.

The generic IoT framework for smart home and building applications proposed in this paper is a comprehensive solution with strengths in integration, flexibility, and advanced functional capabilities. The latter, in particular, requires consideration when implemented in smart-home applications. The underlying integration of BACS and BMS techniques with the IoT poses challenges, including the potential complexity of implementation, the intensity of use of resources available in the network node modules, and data security threats. Therefore, the successful implementation of the generic IoT platform based on the presented framework depends on the effective management of system complexity, tracking technological trends, and solving security and compatibility issues in order to meet the changing needs of the smart home and building industry.

What is very important and significant is to address the weaknesses and threats identified in the SWOT analysis for generic IoT in smart home and building applications; the following research and development directions can be proposed:

Reducing weaknesses

1. Simplify implementation processes by developing automated deployment tools and standardized templates to simplify the installation and configuration of IoT devices in smart homes and buildings. Automation and standardization can minimize the complexity of implementation, making it more user-friendly and reducing the potential for errors;
2. Resource optimization for advanced functions by exploring lightweight algorithms and edge computing strategies to optimize resource-intensive functions, such as machine learning and AI, to ensure efficient operation in resource-constrained environments. Optimizing resources reduces the load on devices and networks, improving overall system performance;
3. Enhance cybersecurity measures by exploring blockchain-based security frameworks, decentralized identity management, and real-time threat detection to strengthen the security posture of smart home and building IoT systems. Implementing advanced cybersecurity measures will strengthen defenses against evolving threats, protect sensitive data, and ensure the integrity of the system.

Mitigating threats

1. Enhance cybersecurity awareness and education by conducting research on effective cybersecurity awareness and education programs for both users and developers involved in IoT applications for smart homes and buildings. Increased awareness and education can empower users to adopt secure practices, reducing the risk of cyber threats such as unauthorized access or data breaches;
2. Standardize security protocols by working with industry stakeholders to establish and promote standardized security protocols for IoT devices and communications in smart home and building ecosystems. Standardization ensures a consistent and robust security framework, making it harder for attackers to exploit vulnerabilities;
3. Continuous monitoring and updating by researching dynamic monitoring solutions and automated update mechanisms to ensure continuous monitoring of IoT systems and rapid deployment of security patches. Proactive monitoring and timely updates reduce the vulnerability window, mitigating potential threats to the IoT ecosystem;
4. Interoperability testing by developing comprehensive interoperability testing frameworks to verify the compatibility of IoT devices with different platforms and protocols. Ensuring interoperability reduces the likelihood of integration challenges and enhances the overall reliability of smart home and building IoT systems.

6. Conclusions

IoT technologies set the direction for the development of many industries related to IT and automation. In particular, in line with the concept of distributed system architecture, they are increasingly entering the structures of BACS networks in smart home and building applications. Along with this process, the technological and functional complexity of these types of systems increases. This paper provides a systematic literature review of the state-of-the-art development of several aspects related to the development of modern smart home and building platforms. The author traced the path of changes in the architecture of distributed automation systems, with an analysis of new edge and fog computing paradigms, implemented at the level of local BACS networks, BMS with IoT modules, and TCP/IP communication channels. Then, application areas for big-data-processing technologies and the implementation of advanced ML and AI techniques supporting the implementation of control functions and effective management of the infrastructure of houses and buildings were identified and discussed. Finally, there is proposed the framework structure for a generic IoT dedicated to applications in building automation in elements of Internet services and local automation servers. A SWOT analysis was performed for the proposed framework in the context of the potential use of BACS network systems with IoT elements in smart home and building applications.

Future research and development work in the generic IoT concept for smart home and building applications could explore enhancing interoperability through standardized communication protocols for seamless integration with a diverse range of devices, for example within platforms like Home Assistant. Moreover, investigating ML applications within Home Assistant and other similar tools can further optimize automation rules, offering personalized and context-aware user experiences. Additionally, exploring energy-efficient algorithms and predictive analytics within the proposed framework could contribute to resource-management efforts and improve overall sustainability in smart homes and buildings.

Funding: This research received no external funding.

Data Availability Statement: Data are contained within the article.

Conflicts of Interest: The author declares no conflicts of interest.

Abbreviations

The following abbreviations are used in this manuscript:

AI	Artificial Intelligence
API	Application Programming Interface
AWS	Amazon Web Services
BaaS	Building as a Service
BACS	Building Automation and Control Systems
BIM	Building Information Modeling
BLE	Bluetooth Low Energy
BMS	Building Management Systems
CoAP	Constrained Application Protocol
DoS	Denial-of-Service
DSM	Demand Side Management
DSR	Demand Side Response
DT	Digital Twin
DTLS	Datagram Transport Layer Security
EPBD	Energy Performance of Buildings Directive
FL	Federated Learning
FM	Facility Management
FoE	Fog of Everything
HVAC	Heating, Ventilation, Air Conditioning
ICT-	Information and Communications Technology
IDS	Intrusion Detection Systems
IoE	Internet of Everything
IoT	Internet of Things
IOTA	Internet of Things Application
IPS	Intrusion Prevention Systems (IPS)
ML	Machine Learning
MQTT	Message Queuing Telemetry Transport protocol
OPC	OLE for Process Control (OLE—Object Linking and Embedding)
P2P	Peer-to-Peer
PLC-	Programmable Logic Controller
RES-	Renewable Energy Sources
SoC-	System-on-a-Chip
SRI-	Smart Readiness Indicator
TIoT	Tactile Internet of Things
TLS	Transport Layer Security
WoT	Web of Things

References

1. Benotmane, M.; Elhari, K.; Kabbaj, A. A Review & Analysis of Current IoT Maturity & Readiness Models and Novel Proposal. *Sci. Afr.* **2023**, *21*, e01748. [CrossRef]
2. Khattak, S.B.A.; Nasralla, M.M.; Farman, H.; Choudhury, N. Performance Evaluation of an IEEE 802.15.4-Based Thread Network for Efficient Internet of Things Communications in Smart Cities. *Appl. Sci.* **2023**, *13*, 7745. [CrossRef]
3. Ferrández-Pastor, F.-J.; Mora, H.; Jimeno-Morenilla, A.; Volckaert, B. Deployment of IoT Edge and Fog Computing Technologies to Develop Smart Building Services. *Sustainability* **2018**, *10*, 3832. [CrossRef]
4. Ali, O.; Ishak, M.K. Bringing Intelligence to IoT Edge: Machine Learning Based Smart City Image Classification Using Microsoft Azure IoT and Custom Vision. *J. Phys. Conf. Ser.* **2020**, *1529*, 042076. [CrossRef]
5. Taghizad-Tavana, K.; Ghanbari-Ghalehjoughi, M.; Razzaghi-Asl, N.; Nojavan, S.; Alizadeh, A. An Overview of the Architecture of Home Energy Management System as Microgrids, Automation Systems, Communication Protocols, Security, and Cyber Challenges. *Sustainability* **2022**, *14*, 15938. [CrossRef]
6. Sharma, H.; Haque, A.; Blaabjerg, F. Machine Learning in Wireless Sensor Networks for Smart Cities: A Survey. *Electronics* **2021**, *10*, 1012. [CrossRef]
7. Wang, B.; Li, M.; Jin, X.; Guo, C. A Reliable IoT Edge Computing Trust Management Mechanism for Smart Cities. *IEEE Access* **2020**, *8*, 46373–46399. [CrossRef]
8. *ISO/IEC 14543-3-10:2020*; Information Technology Home Electronic Systems (HES) Architecture—KNX. International Organization for Standardization: Geneva, Switzerland, 2020.
9. *ISO/IEC 14908-1:2012*; Information Technology Control Network Protocol—LonWorks. International Organization for Standardization: Geneva, Switzerland, 2012.
10. *ISO 16484-6:2020*; Building Automation and Control Systems (BACS)—BACnet. International Organization for Standardization: Geneva, Switzerland, 2020.
11. Bovet, G.; Hennebert, J. Will Web Technologies Impact on Building Automation Systems Architecture? *Procedia Comput. Sci.* **2014**, *32*, 985–990. [CrossRef]
12. Ożadowicz, A. A New Concept of Active Demand Side Management for Energy Efficient Prosumer Microgrids with Smart Building Technologies. *Energies* **2017**, *10*, 1771. [CrossRef]
13. Schraven, M.H.; Droste, K.; Guarnieri Calò Carducci, C.G.C.; Müller, D.; Monti, A. Open-Source Internet of Things Gateways for Building Automation Applications. *J. Sens. Actuator Netw.* **2022**, *11*, 74. [CrossRef]
14. Froiz-Míguez, I.; Fernández-Caramés, T.; Fraga-Lamas, P.; Castedo, L. Design, Implementation and Practical Evaluation of an IoT Home Automation System for Fog Computing Applications Based on MQTT and ZigBee-WiFi Sensor Nodes. *Sensors* **2018**, *18*, 2660. [CrossRef] [PubMed]
15. Petkov, N.; Naumov, A. Overview of Industrial Communication in Process Automation. In Proceedings of the 2022 International Conference Automatics and Informatics (ICAI), Varna, Bulgaria, 6–8 October 2022; IEEE: Piscataway, NJ, USA, 2022; pp. 229–234.
16. Secgin, S. Seven Layers of ISO/OSI. In *Evolution of Wireless Communication Ecosystems*; Wiley: Hoboken, NJ, USA, 2023; pp. 41–50.
17. Vernadat, F.B. Interoperability and Standards for Automation. In *Springer Handbook of Automation*; Nof, S.Y., Ed.; Springer International Publishing: Cham, Switzerland, 2023; pp. 729–752.
18. Kato, T.; Ishikawa, N.; Yoshida, N. Distributed Autonomous Control of Home Appliances Based on Event Driven Architecture. In Proceedings of the 2017 IEEE 6th Global Conference on Consumer Electronics (GCCE), Nagoya, Japan, 24–27 October 2017; IEEE: Piscataway, NJ, USA, 2017; pp. 1–2.
19. Graveto, V.; Cruz, T.; Simöes, P. Security of Building Automation and Control Systems: Survey and Future Research Directions. *Comput. Secur.* **2022**, *112*, 102527. [CrossRef]
20. Martirano, L.; Mitolo, M. Building Automation and Control Systems (BACS): A Review. In Proceedings of the 2020 IEEE International Conference on Environment and Electrical Engineering and 2020 IEEE Industrial and Commercial Power Systems Europe (EEEIC/I&CPS Europe), Madrid, Spain, 9–12 June 2020; IEEE: Piscataway, NJ, USA, 2020; pp. 1–8.
21. Ożadowicz, A.; Grela, J. An Event-Driven Building Energy Management System Enabling Active Demand Side Management. In Proceedings of the 2016 Second International Conference on Event-based Control, Communication, and Signal Processing (EBCCSP), Krakow, Poland, 13–15 June 2016; IEEE: Piscataway, NJ, USA, 2016; pp. 1–8.
22. Pirbhulal, S.; Zhang, H.; E Alahi, M.; Ghayvat, H.; Mukhopadhyay, S.; Zhang, Y.-T.; Wu, W. A Novel Secure IoT-Based Smart Home Automation System Using a Wireless Sensor Network. *Sensors* **2016**, *17*, 69. [CrossRef] [PubMed]
23. Ożadowicz, A. Technical, Qualitative and Energy Analysis of Wireless Control Modules for Distributed Smart Home Systems. *Future Internet* **2023**, *15*, 316. [CrossRef]
24. Prakosa, S.; Nugraha, A.; Atiq, M. Miniature SmartHome Dengan Sonoff. *J. Ris. Rumpun Ilmu Tek.* **2023**, *2*, 41–55.
25. Yang, H.; Kim, B.; Lee, J.; Ahn, Y.; Lee, C. Advanced Wireless Sensor Networks for Sustainable Buildings Using Building Ducts. *Sustainability* **2018**, *10*, 2628. [CrossRef]
26. Anush, K.S.; Sasikala, S.; Arun, K.S.; Arunan, R.; Asfaq, M.A. Enhanced and Secured Smart Home Using Z-Wave Technology. In Proceedings of the 2nd International Conference on Advancements in Electrical, Electronics, Communication, Computing and Automation, ICAECA 2023, Coimbatore, India, 16–17 June 2023; Institute of Electrical and Electronics Engineers Inc.: Piscataway, NJ, USA, 2023.

27. Kazeem, O.O.; Akintade, O.; Kehinde, L.O.; Akintade, O.O.; Kehinde, L.O. Comparative Study of Communication Interfaces for Sensors and Actuators in the Cloud of Internet of Things. *Int. J. Internet Things* **2017**, *2017*, 9–13.
28. Wang, J. Zigbee Light Link and Its Applicationss. *IEEE Wirel. Commun.* **2013**, *20*, 6–7. [CrossRef]
29. Rohini, S.; Venkatasubramanian, K. Z-Wave Based Zoning Sensor for Smart Thermostats. *Indian J. Sci. Technol.* **2015**, *8*, 1–6. [CrossRef]
30. Ali, A.I.; Partal, S.Z.; Kepke, S.; Partal, H.P. ZigBee and LoRa Based Wireless Sensors for Smart Environment and IoT Applications. In Proceedings of the 2019 1st Global Power, Energy and Communication Conference (GPECOM), Nevsehir, Turkey, 12–15 June 2019; IEEE: Piscataway, NJ, USA, 2019; pp. 19–23.
31. Yassein, M.B.; Mardini, W.; Khalil, A. Smart Homes Automation Using Z-Wave Protocol. In Proceedings of the 2016 International Conference on Engineering & MIS (ICEMIS), Agadir, Morocco, 22–24 September 2016; IEEE: Piscataway, NJ, USA, 2016; pp. 1–6.
32. Filippoupolitis, A.; Oliff, W.; Loukas, G. Bluetooth Low Energy Based Occupancy Detection for Emergency Management. In Proceedings of the 2016 15th International Conference on Ubiquitous Computing and Communications and 2016 International Symposium on Cyberspace and Security (IUCC-CSS), Granada, Spain, 14–16 December 2016; IEEE: Piscataway, NJ, USA, 2016; pp. 31–38.
33. Zhuang, Y.; Yang, J.; Li, Y.; Qi, L.; El-Sheimy, N. Smartphone-Based Indoor Localization with Bluetooth Low Energy Beacons. *Sensors* **2016**, *16*, 596. [CrossRef]
34. Collotta, M.; Pau, G. A Solution Based on Bluetooth Low Energy for Smart Home Energy Management. *Energies* **2015**, *8*, 11916–11938. [CrossRef]
35. Tekler, Z.D.; Low, R.; Yuen, C.; Blessing, L. Plug-Mate: An IoT-Based Occupancy-Driven Plug Load Management System in Smart Buildings. *Build. Environ.* **2022**, *223*, 109472. [CrossRef]
36. Balaji, B.; Xu, J.; Nwokafor, A.; Gupta, R.; Agarwal, Y. Sentinel: Occupancy Based HVAC Actuation Using Existing WiFi Infrastructure within Commercial Buildings. In Proceedings of the 11th ACM Conference on Embedded Networked Sensor Systems, Roma, Italy, 11–15 November 2013; ACM: New York, NY, USA, 2013; pp. 1–14.
37. Tekler, Z.D.; Lei, Y.; Dai, X.; Chong, A. Enhancing Personalised Thermal Comfort Models with Active Learning for Improved HVAC Controls. *J. Phys. Conf. Ser.* **2023**, *2600*, 132004. [CrossRef]
38. Singhai, R.; Sushil, R. An Investigation of Various Security and Privacy Issues in Internet of Things. *Mater. Today Proc.* **2023**, *80*, 3393–3397. [CrossRef]
39. Nasir, M.; Muhammad, K.; Ullah, A.; Ahmad, J.; Wook Baik, S.; Sajjad, M. Enabling Automation and Edge Intelligence over Resource Constraint IoT Devices for Smart Home. *Neurocomputing* **2022**, *491*, 494–506. [CrossRef]
40. Lăcătușu, F.; Ionita, A.D.; Lăcătușu, M.; Olteanu, A. Performance Evaluation of Information Gathering from Edge Devices in a Complex of Smart Buildings. *Sensors* **2022**, *22*, 1002. [CrossRef] [PubMed]
41. Babar, M.; Grela, J.; Ożadowicz, A.; Nguyen, P.; Hanzelka, Z.; Kamphuis, I. Energy Flexometer: Transactive Energy-Based Internet of Things Technology. *Energies* **2018**, *11*, 568. [CrossRef]
42. Faqiry, M.; Edmonds, L.; Zhang, H.; Khodaei, A.; Wu, H. Transactive-Market-Based Operation of Distributed Electrical Energy Storage with Grid Constraints. *Energies* **2017**, *10*, 1891. [CrossRef]
43. Pratt, A.; Krishnamurthy, D.; Ruth, M.; Wu, H.; Lunacek, M.; Vaynshenk, P. Transactive Home Energy Management Systems: The Impact of Their Proliferation on the Electric Grid. *IEEE Electrif. Mag.* **2016**, *4*, 8–14. [CrossRef]
44. Ożadowicz, A. A Hybrid Approach in Design of Building Energy Management System with Smart Readiness Indicator and Building as a Service Concept. *Energies* **2022**, *15*, 1432. [CrossRef]
45. Laroui, M.; Nour, B.; Moungla, H.; Cherif, M.A.; Afifi, H.; Guizani, M. Edge and Fog Computing for IoT: A Survey on Current Research Activities & Future Directions. *Comput. Commun.* **2021**, *180*, 210–231. [CrossRef]
46. Yousefpour, A.; Fung, C.; Nguyen, T.; Kadiyala, K.; Jalali, F.; Niakanlahiji, A.; Kong, J.; Jue, J.P. All One Needs to Know about Fog Computing and Related Edge Computing Paradigms: A Complete Survey. *J. Syst. Archit.* **2019**, *98*, 289–330. [CrossRef]
47. Huang, Z.; Lin, K.-J.; Tsai, B.-L.; Yan, S.; Shih, C.-S. Building Edge Intelligence for Online Activity Recognition in Service-Oriented IoT Systems. *Future Gener. Comput. Syst.* **2018**, *87*, 557–567. [CrossRef]
48. Filho, G.P.R.; Meneguette, R.I.; Maia, G.; Pessin, G.; Gonçalves, V.P.; Weigang, L.; Ueyama, J.; Villas, L.A. A Fog-Enabled Smart Home Solution for Decision-Making Using Smart Objects. *Future Gener. Comput. Syst.* **2020**, *103*, 18–27. [CrossRef]
49. Mahmud, R.; Kotagiri, R.; Buyya, R. Fog Computing: A Taxonomy, Survey and Future Directions. In *Internet of Everything: Algorithms, Methodologies, Technologies and Perspectives*; Di Martino, B., Li, K.-C., Yang, L., Esposito, A., Eds.; Springer: Singapore, 2018; pp. 103–130.
50. Alnajar, O.; Barnawi, A. Tactile Internet of Federated Things: Toward Fine-Grained Design of FL-Based Architecture to Meet TIoT Demands. *Comput. Netw.* **2023**, *231*, 109712. [CrossRef]
51. Sun, H.; Yu, H.; Fan, G.; Chen, L. Energy and Time Efficient Task Offloading and Resource Allocation on the Generic IoT-Fog-Cloud Architecture. *Peer Peer Netw. Appl.* **2020**, *13*, 548–563. [CrossRef]
52. Li, W.; Wang, S. A Fully Distributed Optimal Control Approach for Multi-Zone Dedicated Outdoor Air Systems to Be Implemented in IoT-Enabled Building Automation Networks. *Appl. Energy* **2022**, *308*, 118408. [CrossRef]
53. Ge, X.; Yang, F.; Han, Q. Distributed Networked Control Systems: A Brief Overview. *Inf. Sci.* **2017**, *380*, 117–131. [CrossRef]
54. Islam, R.; Rahman, M.W.; Rubaiat, R.; Hasan, M.M.; Reza, M.M.; Rahman, M.M. LoRa and Server-Based Home Automation Using the Internet of Things (IoT). *J. King Saud Univ. Comput. Inf. Sci.* **2022**, *34*, 3703–3712. [CrossRef]

55. Bhatt, J.; Verma, H.K. Design and Development of Wired Building Automation Systems. *Energy Build.* **2015**, *103*, 396–413. [CrossRef]
56. Merz, H.; Hansemann, T.; Hübner, C. *Building Automation*; Signals and Communication Technology; Springer International Publishing: Cham, Switzerland, 2018; ISBN 978-3-319-73222-0.
57. Dietrich, D.; Bruckner, D.; Zucker, G.; Palensky, P. Communication and Computation in Buildings: A Short Introduction and Overview. *IEEE Trans. Ind. Electron.* **2010**, *57*, 3577–3584. [CrossRef]
58. Wang, S. *Intelligent Buildings and Building Automation*; Routledge: London, UK, 2009; ISBN 9781134025107.
59. Domingues, P.; Carreira, P.; Vieira, R.; Kastner, W. Building Automation Systems: Concepts and Technology Review. *Comput. Stand. Interfaces* **2016**, *45*, 1–12. [CrossRef]
60. Lobaccaro, G.; Carlucci, S.; Löfström, E. A Review of Systems and Technologies for Smart Homes and Smart Grids. *Energies* **2016**, *9*, 348. [CrossRef]
61. NORDIC Semiconductor Nordic Semiconductor Delivers Industry-Wide Support for KNX IoT Protocol Following Membership of KNX Association. Available online: https://www.nordicsemi.com/Nordic-news/2023/06/nordic-delivers-industry-wide-support-for-knx-iot-protocol-following-membership-of-knx-association (accessed on 8 December 2023).
62. HAGER Connecting the World of Digital Objects and Services with KNX. Available online: https://assets1.sc.hager.com/turkey/files/IoT-Controller.pdf (accessed on 8 December 2023).
63. EXOR Powerful IoT-Ready Interfaces with KNX Interface. Available online: https://www.exorint.com/corporate/press-release/knx-2020 (accessed on 8 December 2023).
64. Kortuem, G.; Kawsar, F.; Sundramoorthy, V.; Fitton, D. Smart Objects as Building Blocks for the Internet of Things. *IEEE Internet Comput.* **2010**, *14*, 44–51. [CrossRef]
65. Bin, S.; Guiqing, Z.; Shaolin, W.; Dong, W. The Development of Management System for Building Equipment Internet of Things. In Proceedings of the 2011 IEEE 3rd International Conference on Communication Software and Networks, Xi'an, China, 27–29 May 2011; IEEE: Piscataway, NJ, USA, 2011; pp. 423–427.
66. Jarvinen, H.; Litvinov, A.; Vuorimaa, P. Integration Platform for Home and Building Automation Systems. In Proceedings of the 2011 IEEE Consumer Communications and Networking Conference (CCNC), Las Vegas, NV, USA, 9–12 January 2011; IEEE: Piscataway, NJ, USA, 2011; pp. 292–296.
67. Jung, M.; Reinisch, C.; Kastner, W. Integrating Building Automation Systems and IPv6 in the Internet of Things. In Proceedings of the 2012 Sixth International Conference on Innovative Mobile and Internet Services in Ubiquitous Computing, Palermo, Italy, 4–6 July 2012; IEEE: Piscataway, NJ, USA, 2012; pp. 683–688.
68. Lilis, G.; Conus, G.; Asadi, N.; Kayal, M. Towards the next Generation of Intelligent Building: An Assessment Study of Current Automation and Future IoT Based Systems with a Proposal for Transitional Design. *Sustain. Cities Soc.* **2017**, *28*, 473–481. [CrossRef]
69. Lilis, G.; Conus, G.; Kayal, M. A Distributed, Event-Driven Building Management Platform on Web Technologies. In Proceedings of the 2015 International Conference on Event-based Control, Communication, and Signal Processing (EBCCSP), Krakow, Poland, 17–19 June 2015; IEEE: Piscataway, NJ, USA, 2015; pp. 1–8.
70. Conus, G.; Lilis, G.; Zanjani, N.A.; Kayal, M. Toward Event-Driven Mechanism for Load Profile Generation. In Proceedings of the 2017 22nd IEEE International Conference on Emerging Technologies and Factory Automation (ETFA), Limassol, Cyprus, 12–15 September 2017; IEEE: Piscataway, NJ, USA, 2017; pp. 1–6.
71. Mansour, M.; Gamal, A.; Ahmed, A.I.; Said, L.A.; Elbaz, A.; Herencsar, N.; Soltan, A. Internet of Things: A Comprehensive Overview on Protocols, Architectures, Technologies, Simulation Tools, and Future Directions. *Energies* **2023**, *16*, 3465. [CrossRef]
72. Ramprasad, B.; McArthur, J.; Fokaefs, M.; Barna, C.; Damm, M.; Litoiu, M. Leveraging Existing Sensor Networks as IoT Devices for Smart Buildings. In Proceedings of the 2018 IEEE 4th World Forum on Internet of Things (WF-IoT), Singapore, 5–8 February 2018; IEEE: Piscataway, NJ, USA, 2018; pp. 452–457.
73. Vicari, N.; Wuchner, E.; Broring, A.; Niedermeier, C. Engineering and operation made easy—A semantics and ser-vice-oriented approach to building automation. In Proceedings of the 2015 IEEE 20th Conference on Emerging Technologies & Factory Automation (ETFA), Luxembourg, 8–11 September 2015; pp. 1–8. [CrossRef]
74. Baccarelli, E.; Naranjo, P.G.V.; Scarpiniti, M.; Shojafar, M.; Abawajy, J.H. Fog of Everything: Energy-Efficient Networked Computing Architectures, Research Challenges, and a Case Study. *IEEE Access* **2017**, *5*, 9882–9910. [CrossRef]
75. Wildenauer, A.; Mbabu, A.; Underwood, J.; Basl, J. Building-as-a-Service: Theoretical Foundations and Conceptual Framework. *Buildings* **2022**, *12*, 1594. [CrossRef]
76. Wildenauer, A.; Basl, J. Building-as-a-Service: The Opportunities of Service-Dominant Logic for Construction. *Mark. Sci. Inspir.* **2022**, *17*, 41–53. [CrossRef]
77. *European Parliament Directive (EU) 2018/844 of the European Parliament and the Council on the Energy Performance of Buildings*; EU: Strasbourg, France, 2018.
78. Verbeke, S.; Aerts, D.; Reynders, G.; Ma, Y.; Waide, P. *Final Report on the Technical Support to the Development of a Smart Readiness Indicator for Buildings*; European Commission: Brussels, Belgium, 2020.
79. Ramezani, B.; da Silva, M.G.; Simões, N. Application of Smart Readiness Indicator for Mediterranean Buildings in Retrofitting Actions. *Energy Build.* **2021**, *249*, 111173. [CrossRef]

80. Mancini, F.; Lo Basso, G.; de Santoli, L. Energy Use in Residential Buildings: Impact of Building Automation Control Systems on Energy Performance and Flexibility. *Energies* **2019**, *12*, 2896. [CrossRef]
81. Fokaides, P.A.; Panteli, C.; Panayidou, A. How Are the Smart Readiness Indicators Expected to Affect the Energy Performance of Buildings: First Evidence and Perspectives. *Sustainability* **2020**, *12*, 9496. [CrossRef]
82. Vasilopoulos, V.G.; Dimara, A.; Krinidis, S.; Almpanis, P.; Margaritis, N.; Nikolopoulos, N.; Ioannidis, D.; Tzovaras, D. An IoT M2M Architecture for BMS Using Multiple Connectivity Technologies: A Practical Approach. In Proceedings of the 2021 6th International Conference on Smart and Sustainable Technologies (SpliTech), Bol and Split, Croatia, 8–11 September 2021; IEEE: Piscataway, NJ, USA, 2021; pp. 1–6.
83. Fetahi, E.; Ajdari, J.; Zenuni, X.; Hamiti, M. A Cloud Centric Smart City Construction with an IoT Enabled Traffic Prediction Mechanism. In Proceedings of the 2022 11th Mediterranean Conference on Embedded Computing (MECO), Budva, Montenegro, 7–10 June 2022; IEEE: Piscataway, NJ, USA, 2022; pp. 1–6.
84. Chen, Y.-C.; Chang, Y.-C.; Chen, C.-H.; Lin, Y.-S.; Chen, J.-L.; Chang, Y.-Y. Cloud-Fog Computing for Information-Centric Internet-of-Things Applications. In Proceedings of the 2017 International Conference on Applied System Innovation (ICASI), Sapporo, Japan, 13–17 May 2017; IEEE: Piscataway, NJ, USA, 2017; pp. 637–640.
85. Sahil; Sood, S.K. Fog-Cloud Centric IoT-Based Cyber Physical Framework for Panic Oriented Disaster Evacuation in Smart Cities. *Earth Sci. Inform.* **2022**, *15*, 1449–1470. [CrossRef]
86. NVIDIA Developer Jetson Nano Developer Kit. Available online: https://developer.nvidia.com/embedded/jetson-nano-developer-kit (accessed on 14 December 2023).
87. Yar, H.; Imran, A.S.; Khan, Z.A.; Sajjad, M.; Kastrati, Z. Towards Smart Home Automation Using IoT-Enabled Edge-Computing Paradigm. *Sensors* **2021**, *21*, 4932. [CrossRef] [PubMed]
88. Alshaikhli, M.; Elfouly, T.; Elharrouss, O.; Mohamed, A.; Ottakath, N. Evolution of Internet of Things From Blockchain to IOTA: A Survey. *IEEE Access* **2022**, *10*, 844–866. [CrossRef]
89. Parikh, S.; Dave, D.; Patel, R.; Doshi, N. Security and Privacy Issues in Cloud, Fog and Edge Computing. *Procedia Comput. Sci.* **2019**, *160*, 734–739. [CrossRef]
90. Alwakeel, A.M. An Overview of Fog Computing and Edge Computing Security and Privacy Issues. *Sensors* **2021**, *21*, 8226. [CrossRef]
91. Ullah, A.; Ullah, S.I.; Salam, A. Internal DoS Attack Detection and Prevention in Fog Computing. In Proceedings of the 2021 International Conference on Information Technology (ICIT), Amman, Jordan, 14–15 July 2021; IEEE: Piscataway, NJ, USA, 2021; pp. 763–768.
92. Huso, I.; Piro, G.; Boggia, G. Distributed and Privacy-Preserving Data Dissemination at the Network Edge via Attribute-Based Searchable Encryption. In Proceedings of the 2022 20th Mediterranean Communication and Computer Networking Conference (MedComNet), Pafos, Cyprus, 1–3 June 2022; IEEE: Piscataway, NJ, USA, 2022; pp. 122–130.
93. Xu, R.; Palanisamy, B.; Joshi, J. QueryGuard: Privacy-Preserving Latency-Aware Query Optimization for Edge Computing. In Proceedings of the 2018 17th IEEE International Conference on Trust, Security and Privacy in Computing and Communications/12th IEEE International Conference on Big Data Science and Engineering (TrustCom/BigDataSE), New York, NY, USA, 1–3 August 2018; IEEE: Piscataway, NJ, USA, 2018; pp. 1097–1106.
94. Alam, T. Blockchain-Based Internet of Things: Review, Current Trends, Applications, and Future Challenges. *Computers* **2022**, *12*, 6. [CrossRef]
95. Moniruzzaman, M.; Khezr, S.; Yassine, A.; Benlamri, R. Blockchain for Smart Homes: Review of Current Trends and Research Challenges. *Comput. Electr. Eng.* **2020**, *83*, 106585. [CrossRef]
96. Taneja, S.; Rana, I.; Tyagi, S. Blockchain for Iot Security and Privacy: Smart Home—A Review. *Int. J. Adv. Eng. Manag.* **2023**, *5*, 1293.
97. Afzal, M.; Huang, Q.; Amin, W.; Umer, K.; Raza, A.; Naeem, M. Blockchain Enabled Distributed Demand Side Management in Community Energy System with Smart Homes. *IEEE Access* **2020**, *8*, 37428–37439. [CrossRef]
98. Tyagi, A.K.; Dananjayan, S.; Agarwal, D.; Thariq Ahmed, H.F. Blockchain—Internet of Things Applications: Opportunities and Challenges for Industry 4.0 and Society 5.0. *Sensors* **2023**, *23*, 947. [CrossRef]
99. Arif, S.; Khan, M.A.; Rehman, S.U.; Kabir, M.A.; Imran, M. Investigating Smart Home Security: Is Blockchain the Answer? *IEEE Access* **2020**, *8*, 117802–117816. [CrossRef]
100. Abed, S.; Jaffal, R.; Mohd, B.J. A Review on Blockchain and IoT Integration from Energy, Security and Hardware Perspectives. *Wirel. Pers. Commun.* **2023**, *129*, 2079–2122. [CrossRef]
101. Djenouri, D.; Laidi, R.; Djenouri, Y.; Balasingham, I. Machine Learning for Smart Building Applications. *ACM Comput. Surv.* **2020**, *52*, 24. [CrossRef]
102. Shayeghi, H.; Shahryari, E.; Moradzadeh, M.; Siano, P. A Survey on Microgrid Energy Management Considering Flexible Energy Sources. *Energies* **2019**, *12*, 2156. [CrossRef]
103. Hou, P.; Yang, G.; Hu, J.; Douglass, P.J.; Xue, Y. A Distributed Transactive Energy Mechanism for Integrating PV and Storage Prosumers in Market Operation. *Engineering* **2022**, *12*, 171–182. [CrossRef]
104. Pipattanasomporn, M.; Kuzlu, M.; Rahman, S.; Teklu, Y. Load Profiles of Selected Major Household Appliances and Their Demand Response Opportunities. *IEEE Trans. Smart Grid* **2014**, *5*, 742–750. [CrossRef]

105. Bouchabou, D.; Nguyen, S.M.; Lohr, C.; LeDuc, B.; Kanellos, I. A Survey of Human Activity Recognition in Smart Homes Based on IoT Sensors Algorithms: Taxonomies, Challenges, and Opportunities with Deep Learning. *Sensors* **2021**, *21*, 6037. [CrossRef]
106. Suman, S.; Etemad, A.; Rivest, F. Potential Impacts of Smart Homes on Human Behavior: A Reinforcement Learning Approach. *IEEE Trans. Artif. Intell.* **2022**, *3*, 567–580. [CrossRef]
107. Machorro-Cano, I.; Alor-Hernández, G.; Paredes-Valverde, M.A.; Rodríguez-Mazahua, L.; Sánchez-Cervantes, J.L.; Olmedo-Aguirre, J.O. HEMS-IoT: A Big Data and Machine Learning-Based Smart Home System for Energy Saving. *Energies* **2020**, *13*, 1097. [CrossRef]
108. Shah, S.; Iqbal, M.; Aziz, Z.; Rana, T.; Khalid, A.; Cheah, Y.-N.; Arif, M. The Role of Machine Learning and the Internet of Things in Smart Buildings for Energy Efficiency. *Appl. Sci.* **2022**, *12*, 7882. [CrossRef]
109. Kawa, B.; Borkowski, P. Integration of Machine Learning Solutions in the Building Automation System. *Energies* **2023**, *16*, 4504. [CrossRef]
110. Vassiliades, C.; Agathokleous, R.; Barone, G.; Forzano, C.; Giuzio, G.F.; Palombo, A.; Buonomano, A.; Kalogirou, S. Building Integration of Active Solar Energy Systems: A Review of Geometrical and Architectural Characteristics. *Renew. Sustain. Energy Rev.* **2022**, *164*, 112482. [CrossRef]
111. Fambri, G.; Badami, M.; Tsagkrasoulis, D.; Katsiki, V.; Giannakis, G.; Papanikolaou, A. Demand Flexibility Enabled by Virtual Energy Storage to Improve Renewable Energy Penetration. *Energies* **2020**, *13*, 5128. [CrossRef]
112. Cvitić, I.; Peraković, D.; Periša, M.; Gupta, B. Ensemble Machine Learning Approach for Classification of IoT Devices in Smart Home. *Int. J. Mach. Learn. Cybern.* **2021**, *12*, 3179–3202. [CrossRef]
113. Zhang, C.; Xie, Y.; Bai, H.; Yu, B.; Li, W.; Gao, Y. A Survey on Federated Learning. *Knowl.-Based Syst.* **2021**, *216*, 106775. [CrossRef]
114. Mammen, P.M. Federated Learning: Opportunities and Challenges. *arXiv* **2021**, arXiv:2101.05428.
115. Zubaydi, H.D.; Varga, P.; Molnár, S. Leveraging Blockchain Technology for Ensuring Security and Privacy Aspects in Internet of Things: A Systematic Literature Review. *Sensors* **2023**, *23*, 788. [CrossRef]
116. Yazdinejad, A.; Dehghantanha, A.; Parizi, R.M.; Srivastava, G.; Karimipour, H. Secure Intelligent Fuzzy Blockchain Framework: Effective Threat Detection in IoT Networks. *Comput. Ind.* **2023**, *144*, 103801. [CrossRef]
117. Khan, M.A.; Abbas, S.; Rehman, A.; Saeed, Y.; Zeb, A.; Uddin, M.I.; Nasser, N.; Ali, A. A Machine Learning Approach for Blockchain-Based Smart Home Networks Security. *IEEE Netw.* **2021**, *35*, 223–229. [CrossRef]
118. Panchalingam, R.; Chan, K.C. A State-of-the-Art Review on Artificial Intelligence for Smart Buildings. *Intell. Build. Int.* **2021**, *13*, 203–226. [CrossRef]
119. Rodriguez-Garcia, P.; Li, Y.; Lopez-Lopez, D.; Juan, A.A. Strategic Decision Making in Smart Home Ecosystems: A Review on the Use of Artificial Intelligence and Internet of Things. *Internet Things* **2023**, *22*, 100772. [CrossRef]
120. Genkin, M.; McArthur, J.J. B-SMART: A Reference Architecture for Artificially Intelligent Autonomic Smart Buildings. *Eng. Appl. Artif. Intell.* **2023**, *121*, 106063. [CrossRef]
121. Hou, Z.; She, C.; Li, Y.; Niyato, D.; Dohler, M.; Vucetic, B. Intelligent Communications for Tactile Internet in 6G: Requirements, Technologies, and Challenges. *IEEE Commun. Mag.* **2021**, *59*, 82–88. [CrossRef]
122. Fanibhare, V.; Sarkar, N.I.; Al-Anbuky, A. A Survey of the Tactile Internet: Design Issues and Challenges, Applications, and Future Directions. *Electronics* **2021**, *10*, 2171. [CrossRef]
123. Akshatha, N.; Rai, K.H.; Haritha, M.K.; Ramesh, R.; Hegde, R.; Kumar, S. Tactile Internet: Next Generation IoT. In Proceedings of the 2019 Third International Conference on Inventive Systems and Control (ICISC), Coimbatore, India, 10–11 January 2019; IEEE: Piscataway, NJ, USA, 2019; pp. 22–26.
124. Promwongsa, N.; Ebrahimzadeh, A.; Naboulsi, D.; Kianpisheh, S.; Belqasmi, F.; Glitho, R.; Crespi, N.; Alfandi, O. A Comprehensive Survey of the Tactile Internet: State-of-the-Art and Research Directions. *IEEE Commun. Surv. Tutor.* **2021**, *23*, 472–523. [CrossRef]
125. Eneyew, D.D.; Capretz, M.A.M.; Bitsuamlak, G.T. Toward Smart-Building Digital Twins: BIM and IoT Data Integration. *IEEE Access* **2022**, *10*, 130487–130506. [CrossRef]
126. Coupry, C.; Noblecourt, S.; Richard, P.; Baudry, D.; Bigaud, D. BIM-Based Digital Twin and XR Devices to Improve Maintenance Procedures in Smart Buildings: A Literature Review. *Appl. Sci.* **2021**, *11*, 6810. [CrossRef]
127. Hadjidemetriou, L.; Stylianidis, N.; Englezos, D.; Papadopoulos, P.; Eliades, D.; Timotheou, S.; Polycarpou, M.M.; Panayiotou, C. A Digital Twin Architecture for Real-Time and Offline High Granularity Analysis in Smart Buildings. *Sustain. Cities Soc.* **2023**, *98*, 104795. [CrossRef]
128. Chiariotti, F.; Condoluci, M.; Mahmoodi, T.; Zanella, A. SymbioCity: Smart Cities for Smarter Networks. *Trans. Emerg. Telecommun. Technol.* **2018**, *29*, e3206. [CrossRef]
129. Sarathchandra, C.; Robitzsch, S.; Ghassemian, M.; Olvera-Hernandez, U. Enabling Bi-Directional Haptic Control in Next Generation Communication Systems: Research, Standards, and Vision. In Proceedings of the 2021 IEEE Conference on Standards for Communications and Networking (CSCN), Thessaloniki, Greece, 15–17 December 2021; IEEE: Piscataway, NJ, USA, 2021; pp. 99–104.
130. Gupta, R.; Tanwar, S.; Tyagi, S.; Kumar, N. Tactile Internet and Its Applications in 5G Era: A Comprehensive Review. *Int. J. Commun. Syst.* **2019**, *32*, e3981. [CrossRef]
131. Mekikis, P.-V.; Ramantas, K.; Antonopoulos, A.; Kartsakli, E.; Sanabria-Russo, L.; Serra, J.; Pubill, D.; Verikoukis, C. NFV-Enabled Experimental Platform for 5G Tactile Internet Support in Industrial Environments. *IEEE Trans. Ind. Inform.* **2020**, *16*, 1895–1903. [CrossRef]

132. Baghalzadeh Shishehgarkhaneh, M.; Keivani, A.; Moehler, R.C.; Jelodari, N.; Roshdi Laleh, S. Internet of Things (IoT), Building Information Modeling (BIM), and Digital Twin (DT) in Construction Industry: A Review, Bibliometric, and Network Analysis. *Buildings* **2022**, *12*, 1503. [CrossRef]
133. Khajavi, S.H.; Motlagh, N.H.; Jaribion, A.; Werner, L.C.; Holmstrom, J. Digital Twin: Vision, Benefits, Boundaries, and Creation for Buildings. *IEEE Access* **2019**, *7*, 147406–147419. [CrossRef]
134. Arsenijević, D.; Stankovski, S.; Ostojić, G.; Baranovski, I.; Oros, D. An Overview of IoT Readiness Assessment Methods. In Proceedings of the 8th International Conference on Information Society and Technology, Kopaonik, Serbia, 11–14 March 2018; pp. 48–53.
135. Metwally, E.A.; Farid, A.A.; Ismail, M.R. Development of an IoT Assessment Method: An Interdisciplinary Framework for Energy Efficient Buildings. *Energy Build.* **2022**, *254*, 111545. [CrossRef]
136. *EN 15232*; European Committee for Standarization Energy Performance of Buildings—Impact of Building Automation, Controls and Building Management. European Commission: Brussels, Belgium, 2017.
137. *ISO 52120-1:2021*; I. 205 T.C. Energy Performance of Buildings Contribution of Building Automation, Controls and Building Management. International Organization for Standardization: Geneva, Switzerland, 2021.
138. Wei, W.; Lee, K.; Murray, D. Building a Generic Architecture for the Internet of Things. In Proceedings of the 2013 IEEE Eighth International Conference on Intelligent Sensors, Sensor Networks and Information Processing, Melbourne, VIC, Australia, 2–5 April 2013; IEEE: Piscataway, NJ, USA, 2013; pp. 333–338.
139. Wang, W.; Lee, K.; Murray, D. A Global Generic Architecture for the Future Internet of Things. *Serv. Oriented Comput. Appl.* **2017**, *11*, 329–344. [CrossRef]
140. Ali, Z.; Mahmood, A.; Khatoon, S.; Alhakami, W.; Ullah, S.S.; Iqbal, J.; Hussain, S. A Generic Internet of Things (IoT) Middleware for Smart City Applications. *Sustainability* **2022**, *15*, 743. [CrossRef]

computers

MDPI

Article

Classifying the Main Technology Clusters and Assignees of Home Automation Networks Using Patent Classifications

Konstantinos Charmanas [1], Konstantinos Georgiou [1], Nikolaos Mittas [2] and Lefteris Angelis [1,*]

[1] School of Informatics, Aristotle University of Thessaloniki, 54124 Thessaloniki, Greece; kcharman@csd.auth.gr (K.C.); konsgeor@csd.auth.gr (K.G.)
[2] Department of Chemistry, International Hellenic University, 65404 Kavala, Greece; nmittas@chem.ihu.gr
* Correspondence: lef@csd.auth.gr

Abstract: Home automation technologies are a vital part of humanity, as they provide convenience in otherwise mundane and repetitive tasks. In recent years, given the development of the Internet of Things (IoT) and artificial intelligence (AI) sectors, these technologies have seen a tremendous rise, both in the methodologies utilized and in their industrial impact. Hence, many organizations and companies are securing commercial rights by patenting such technologies. In this study, we employ an analysis of 8482 home automation patents from the United States Patent and Trademark Office (USPTO) to extract thematic clusters and distinguish those that drive the market and those that have declined over the course of time. Moreover, we identify prevalent competitors per cluster and analyze the results under the spectrum of their market impact and objectives. The key findings indicate that home automation networks encompass a variety of technological areas and organizations with diverse interests.

Keywords: patent analysis; home automation networks; patent classifications; cluster analysis; technology forecasting; competitor analysis

Citation: Charmanas, K.; Georgiou, K.; Mittas, N.; Angelis, L. Classifying the Main Technology Clusters and Assignees of Home Automation Networks Using Patent Classifications. *Computers* **2023**, *12*, 211. https://doi.org/10.3390/computers12100211

Academic Editor: Sergio Correia

Received: 21 September 2023
Revised: 16 October 2023
Accepted: 18 October 2023
Published: 20 October 2023

1. Introduction

Home automation systems and networks aim to facilitate communication between devices by integrating technologies that accommodate automated procedures. The many applications of home automation networks leverage network technologies, sensors, controllers and devices to establish systems of interconnected devices and networks, e.g., security systems, home appliances, energy management systems and multiple similar architectures [1]. According to Sovacool and Del [2], home automation technologies offer many benefits that are related to finance, healthcare, security, education and entertainment, thus affecting several aspects of both daily life and industrial services. From the user perspective, the main perceptible and important benefits of home automation networks are related to (i) comfort (smart kitchen, TV), (ii) monitoring (sensors), (iii) therapy (automated delivery of treatments) (iv) support (robotic devices, mobility devices) and (v) consultancy (sensors) [3]. This fact, combined with the impressive rise of IoT and AI in the last decade, renders these technologies as catalysts in the progress and development of humanity, hence offering practical services and applications.

In order to provide novel solutions and produce competitive products, both researchers and industries are seeking to solve important problems and optimize the current approaches for innovative home automation technologies by employing a variety of scientific approaches and models, e.g., deep learning, statistics, graph theory, cryptography, heuristic algorithms [4]. These innovations, which can potentially lead to commercially exploitable outcomes, are usually covered by patent documents that protect the valuable legal and practical assets of individuals and organizations against competitors.

Regarding the industrial involvement, patented technologies can be used as a basis upon which to assess emerging technologies and key individuals, organizations and countries as they contain significant information and details on both trending and essential technologies and methodologies [5]. The process of analyzing and extracting knowledge from patent data constitutes the field of patent analysis and is used as a tool to cover multiple research goals related to trending and competitor analysis, technology forecasting and strategic planning. Evidently, the corresponding approaches leverage several data and text mining techniques for the completion of defined research goals [6], including natural language processing, cluster analysis and citation networks/graph theory.

In this study, we focus on exploring and assessing the main technologies and assignees of home automation networking by collecting and analyzing relevant patent data in order to provide sufficient insights into emerging and dominant technologies as well as leading assignees. Thus, we first collect the appropriate patent information from the USPTO (https://www.uspto.gov/, accessed on 16 October 2023) and then deploy a patent analysis framework that combines cluster analysis with multiple patent properties to address these objectives. Our framework is inspired by existing patent analysis methodologies that make use of patent classification schemas [7,8], in our case the Cooperative Patent Classification (CPC (https://www.uspto.gov/web/patents/classification/cpc/html/cpc.html), accessed on 16 October 2023), and the Compound Annual Growth (CAGR) [9], to detect dominant or declining clusters. In brief, we firstly establish a CPC subclass to a patent matrix (CPM) and further employe the non-negative matrix factorization (NMF) algorithm [10,11] to detect the main technology clusters of home automation patents. Additionally, we leverage additional information from the patent data to firstly calculate the CAGR of the extracted clusters and to evaluate the status and the representative organizations of each cluster.

Overall, our analysis demystifies the underlying relationships between patent classifications which leads to a determination of the general technology clusters of home automation networking patents. The corresponding findings, which are complemented by the evaluation of the CAGR, and the leading assignees of each cluster can provide guidelines for multiple patent analysis scopes such as trending and competitor analysis as well as technology forecasting.

The rest of this paper is organized as follows: in Section 2, we provide a comprehensive review of the related literature both in home automation networks and in patent analysis, hence profiling the progress in the field. In Section 3 we present the utilized methodology along with the key concepts and definitions. In Section 4, the results of our analysis are presented and visualized while Section 5 includes a contained discussion of our main findings. Finally, Sections 6–8 provide the limitations of our study, conclusions and some interesting future work directions.

2. Literature Review

2.1. Home Automation Networks

Home automation networks have recently grown rapidly and are now widely applied to improve different systems and appliances. Shuhaiber and Mashal [12] and Sin et al. [13] have revealed that the perceived usefulness and risks as well as the ease of use of these technologies are some of the most important factors towards accepting and using home automation networks. However, the efficiency and quality of home automation networks is measured by different indicators. According to Toschi et al. [14], home automation technologies should be evaluated based on their characteristics, such as communication and data type; their performance, such as complexity, rate, and processing power; as well as their various expenses, such as cost, energy consumption.

Zielonka et al. [4] focus their review on research reports and patents in order to identify the research trends of home automation applications. According to their findings, the popular and trending technologies aim to improve healthcare, e.g., eldercare; information security, e.g., cryptography and the blockchain; and energy systems, e.g., management of energy consumption. Other areas of interest include remote devices, communication systems and sensors.

The recent advances of machine learning and deep learning architectures have brought to the surface some new technologies and frameworks that employ such techniques in home automation services. Yu et al. [15] discuss the potential usage of deep learning in multiple applications and objectives that are relevant to home automation networks, indicating that these techniques improve existing machine learning approaches, e.g., naive Bayes and support vector machines. Their analysis shows that these applications are related to activity recognition and prediction, security as well as energy management while the utilized data structures are associated with sensors, images, videos and audio.

2.2. Patent Analaysis

Researchers and organizations have acknowledged the value of patent analysis as the information included in patent documents represents an overview of the technologies that are developed for different domains and objectives. The existing research, i.e., patent analysis studies, covers a widespread area and different fields of interest, including electrical vehicles [16], artificial intelligence [17], security [18], software development [19], etc. In general, a patent record contains information concerning patent assignees, usually large companies; inventors; citations; descriptions, i.e., titles and abstracts; and patent classifications, i.e., specific categories and identifiers describing relevant technological fields.

In particular, patent classifications have been effectively used to extract comprehensive knowledge from domain specific datasets. Jee et al. [20] have leveraged the available patent classifications, which are assigned to each patent, in order to identify promising technologies by six different perspectives, indicating their role in a technology area expressed by a representative patent dataset. In a different context, Park and Geum [21] assessed the relationships of the different technology areas using patent data and classifications, with a further goal of identifying potential opportunities from convergence networks. Similarly, Geum and Kim [22] have combined the information from patent classifications to establish a graph network to uncover core technologies and technological chances through cluster analysis. In addition, clustering applications that are based on patent classifications have been previously proposed as effective approaches in forecasting and evaluating promising and emerging technologies [7,8].

Regarding patent assignees, analyzing the patents and assets of a specific company can lead to the evaluation of its overall knowledge status, immediate competitors and relative strategic positioning. Suominen et al. [23] have explored the technologies of telecommunication industries through patent analysis in order to profile the different involved organizations and assess the potential connections between them. Likewise, by analyzing patent citations and the main properties of patent data, Daim et al. [24] evaluated the overall technology knowledge status of the different organizations, thus distinguishing those that hold significant inventions in the field of IoT, cybersecurity and blockchain. Additionally, Wang and Hsu [25] have established a topic strategy matrix and a topic–firm network to assess the activity and associations between topics and firms and to discover the core assignees and significant topics of smart manufacturing technologies.

3. Methodology

The outline of the proposed approach is presented in Figure 1 and consists of distinct phases which are described as (i) data collection, (ii) identification of technology clusters, (iii) evaluation of technology clusters, and (iv) assessment of leading assignees. In addition, we provide an additional figure (Figure 2) presenting the data exchanges between these phases, with the displayed data structures described later in this section. Overall, through the techniques that are described in this section we aim to provide answers to the following research questions (RQs):

RQ_1: *What are the core technology clusters of home automation networks?*

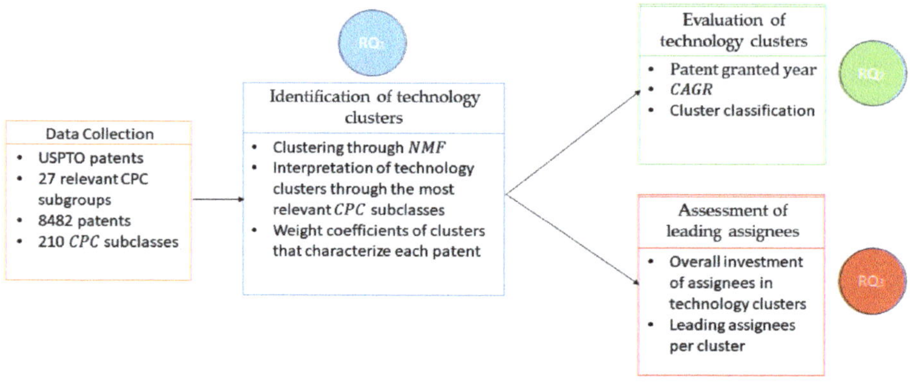

Figure 1. Main outline of the proposed approach.

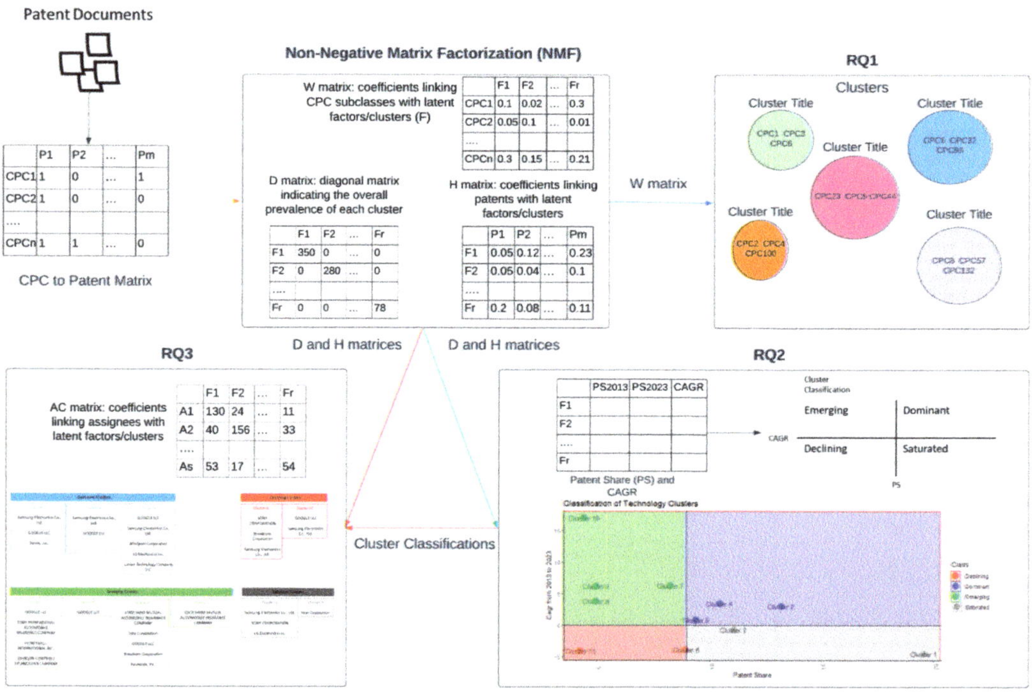

Figure 2. Data exchanges between the phases of the proposed approach.

Purpose: Given that patent data correspond to multifaceted and innovative products that are patented by large companies, it is expected that the technological objectives will be varied and different. Hence, the purpose of RQ_1 is to aggregate the patent data and detect primary clusters of co-occurring technologies that describe the status of home automation networks. These clusters represent groups of CPC subclasses that have similar objectives and are hence related in terms of technological convergence.

RQ_2: *How are the different technology clusters classified with respect to patent activity trends and growth?*

Purpose: The identified technology clusters are composed of patents that cover a wide time period. Thus, it is of high research interest to examine which clusters evolve and dominate the market over time, as this indicates essential technologies. It is also of interest to examine which clusters present a downward trend, indicating declining or niche domains of home automation networks that require more careful planning and more thorough methodologies.

RQ_3: *Who are the leading assignees/competitors of the different technology clusters of home automation networking?*

Purpose: Patent data contain a multitude of information, with one of the most important fields being the patent assignee, i.e., the company, organization or individual, that has the ownership of the patent. By utilizing the technology clusters identified in RQ_1 and the growth of each cluster from RQ_2,, in RQ_3 we are able to pinpoint the leading companies of home automation networking that participate in dominating and emerging clusters while, at the same time, we discuss companies that operate in less popular or niche fields.

The purpose of the first phase of our approach was to identify an appropriate source of patent data, from an acclaimed patent office. Given that USPTO is hailed in similar literature as a potent repository of patent grants, we turned our attention to this specific patent office to accumulate patent data about home automation networks. In addition, as the usage of a manually formulated search string could potentially lead to data omission or the retrieval of irrelevant patents, we decided to leverage one of the existing classification schemas of USPTO, i.e., the CPC schema, and identify all of the CPC subgroups that are relevant to the area of interest of this study, i.e., home automation networks. As a result, we manually searched the CPC schema and identified 27 subgroups, summarized into 9 CPC prefixes, which are presented in more detail in our prior work [26]. Instead of analyzing a sample, which was the objective of our previous work, in this study we analyze the complete dataset that is formed by the 8482 patents.

The CPC schema, used by the USPTO for patent classification, follows a tree structure and is divided into upper levels that correspond to more general and inclusive classes, these are then dispersed into secondary and tertiary levels that reflect more specific technological objectives and domains. In general, a single USPTO patent is assigned, by applicants or examiners, to multiple CPC subgroups that are used to describe the concepts of the respective patent and cover a very specific technological area. As the goal of this study is to identify the general technology clusters of home automation networking patents, we used an upper level of the CPC schema, i.e., the subclasses, instead of the lowest level (subgroups) to achieve our goals. In Figure 3, we present the meaning of the different encodings that form a single CPC subgroup.

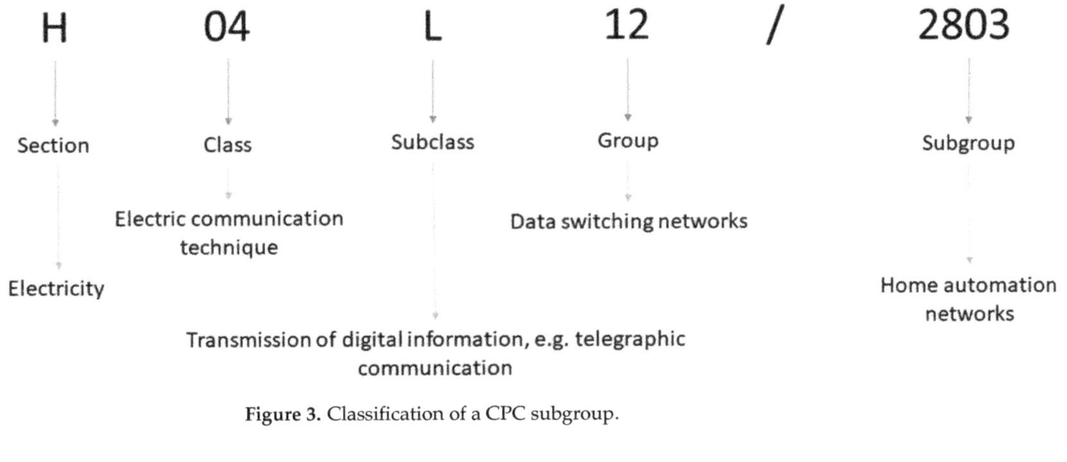

Figure 3. Classification of a CPC subgroup.

After collecting all the necessary patent information, we then proceed to constructing the CPC-to-patent matrix (CPM)

$$CPM_{i,j} = \left\{ \begin{array}{c} 1 \; when \; the \; j-th \; patent \; is \; assigned \; to \; the \; i-th \\ CPC \; subclass \; at \; least \; once \\ 0 \; otherwise \end{array} \right\}, \quad \text{for } i = 1,\, 2, \ldots, n \text{ and } j = 1,\, 2, \ldots, m \tag{1}$$

where m is the number of patents and n is the number of the detected CPC subclasses (210 overall in our dataset).

The CPM will be used to establish cohesive technology clusters, reduce the dimensionality of the CPC schema and identify the potential interconnections between the CPC subclasses, and possibly subgroups, which fall under different encodings. To identify and assess the general technology clusters associated with home automation networking, we employ a variation of the standard NMF [10,11], proposed by Debruine et al. [27],

$$V_{n\times m} = w_{n\times r} d_{r\times r} h_{r\times m} \tag{2}$$

where an initial matrix V (from visible), the CPM in our case, is decomposed into two lower-rank non-negative matrices entitled w and h, for a predefined number of features, which is referred to as rank (r). These two outcoming matrices (w and h) help us relate the CPC subclasses and the patents with their respective semantic features. In this variation, d is a diagonal matrix that scales the two aforementioned matrices to sum to one, i.e., the elementwise column sums of w and the elementwise row sums of h are equal to one and indicate the overall prevalence of each feature in the data.

In general, each column of w defines the vector of a latent/basis feature of V, while a column of h stores the weight coefficients that connect an initial observation with these basis features [28]. Lee and Seung [10] describe how, for the initial matrix $V_{n\times m}$, with n words and m documents, $w_{i,j}$ indicates the frequency of the ith word in the jth semantic feature, while $h_{i,j}$ denotes the weight that is given to the ith semantic feature for the jth document. In our case, where words are replaced by CPC subclasses and documents by patents, w will help us address the core concepts of the basis features, expressing technology clusters, while h will help us assess the overall presence of these features in each patent.

Although NMF is quite an old algorithm, it is considered an effective and widespread methodology as it is still employed in multiple areas of interest, such as sound event detection [29], speech recognition [30], text mining [31–33], image analysis [34,35], security and privacy [36] and community detection [37]. In many of these applications, NMF is used as a tool for classification, filtering, dimensionality reduction as well as data clustering, which is the core concept of this study. In addition, the advantage of NMF against the standard clustering approaches is that this algorithm produces weight coefficients instead of an evaluation that relates each data point/observation to a single cluster.

To pick the 'optimal' rank, we employ normalized pointwise mutual information [38] for the ten CPC subclasses with the highest frequencies in each feature, which is a typical selection to measure the coherence of clusters and topic models. NPMI is a measure, ranging from -1 to 1, for pairwise associations where higher values indicate positive degrees of co-occurrence/association while lower values indicate positive degrees of independence. After finalizing the model for the 'best' number of features, we make use of the most frequent CPC subclasses to interpret each feature into a general technology cluster.

Moreover, we follow the methodology proposed by Choi and Song [9] to classify each cluster as emerging, dominant, declining or saturated by evaluating the compound annual growth (CAGR), given a period of ten years (2013–2023) and the patent share (PS) of the clusters in the collected patents. In this study, we denote the PS as the overall prevalence of each factor/cluster in the patents and l as the patents belonging to a selected patent subset.

$$PS_{f,l} = \frac{\sum_{i \epsilon l} d_{f,f} h_{f,i}}{\sum_{k=1}^{r} \sum_{j \epsilon l} d_{k,k} h_{k,j}}, \text{ for } f = 1, 2, \ldots, r \tag{3}$$

$$CAGR_i = \left(\frac{PS_{i,Until-2023}}{PS_{i,Until-2013}} \right)^{\frac{1}{2023-2013}} - 1 \tag{4}$$

To characterize the clusters into the four classes, we plot the overall PS with the CAGR evaluations, where the characteristics of each class are the following:

- Emerging: Low PS, high CAGR
- Dominant: High PS, high CAGR
- Declining: Low PS, low CAGR
- Saturated: High PS, low CAGR

As one of the goals of this study is to identify the leading assignees of the various technologies that are relevant to home automation networking, we further assess the overall investment of each assignee in a technology cluster using the h matrix

$$AC_{i,j} = \sum_{p \epsilon P_i} d_{j,j} h_{j,p}, \text{ for } i = 1, 2, \ldots, s \text{ and } j = 1, 2, \ldots, r \tag{5}$$

where $AC_{i,j}$ (assignee–cluster matrix) denotes the overall investment of the ith assignee in the jth cluster, s is the number of assignees and P_i is the set of patents which are assigned to the ith assignee. The assignees with the highest AC values in a cluster are declared as the leading assignees.

4. Results

In this section, we present the main results of this study, with which we aim to provide answers to the three posed research questions. To answer RQ_1, we first present the main technology clusters as extracted using the NMF, complemented by representative descriptions, which in turn are based on the most frequent CPC subclasses of each cluster (Section 4.1). Furthermore, in Section 4.2, we discuss the findings concerning the classification of each technology cluster based on CAGR and PS. Finally, in Section 4.3 we present the leading assignees (organizations) of each technology cluster and also evaluate their investment in home automation networking technologies.

4.1. RQ_1: *What Are the Core Technology Clusters of Home Automation Networks?*

Initially, we evaluated 19 NMF models, using the NPMI, ranging from 2 to 20 features/clusters. The evaluations of these models are presented in Figure 4, where we observe that the NPMI is maximized under 11 features/clusters. Thus, we proceed to the following phases using the properties of the respective model.

An initial step towards the interpretation of the derived results is to inspect the most frequent CPC subclasses of each cluster to provide a representative description for each one. These subclasses, along with their relative frequency are presented in Table 1, where we should note that the frequencies of all detected subclasses, not only the top 10, are scaled and are summed to one for each cluster.

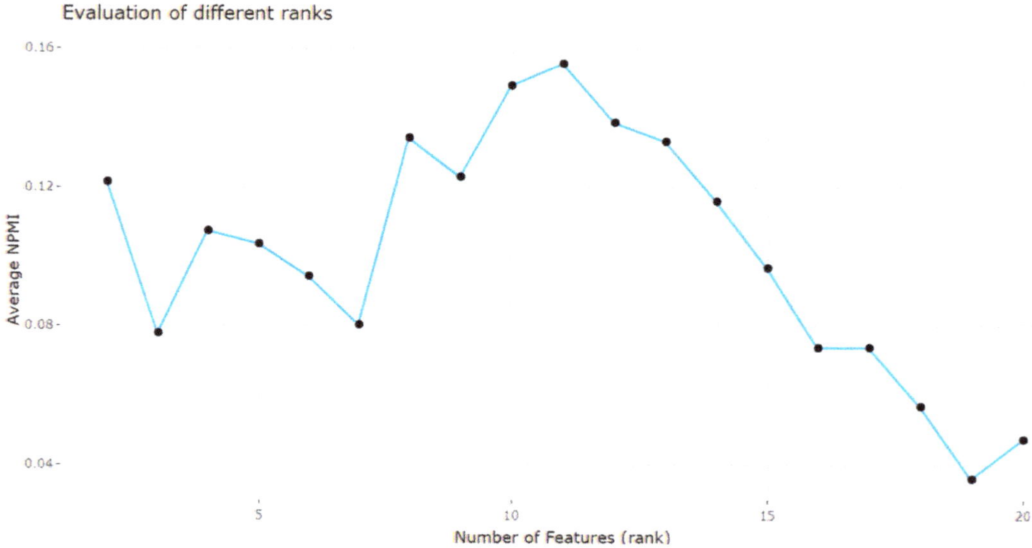

Figure 4. Evaluation of different ranks in NMF.

Table 1. Most frequent CPC subclasses of each cluster.

Cluster 1	Cluster 2	Cluster 3	Cluster 4	Cluster 5	Cluster 6
H04L 0.86453	G06F 0.53989	H04B 0.13629	H04W 0.72243	Y02B 0.24044	H04N 0.59306
H04B 0.02576	H04L 0.28227	G08C 0.12658	H04L 0.1347	Y04S 0.1797	H04L 0.20418
H05B 0.01968	G10L 0.06546	Y02D 0.06945	Y02D 0.05499	H02J 0.12335	G08C 0.02632
G06N 0.00734	H04R 0.02227	H04Q 0.06724	G06K 0.01669	H05B 0.09401	H04H 0.02269
E06B 0.00647	G09G 0.00912	Y10S 0.05167	A63F 0.00776	Y02D 0.05959	G09G 0.02036
H04R 0.00504	G06N 0.00851	H04J 0.049	G09B 0.00709	Y02E 0.0357	G11B 0.01854
F21V 0.00448	G06K 0.00827	Y02P 0.04794	B64C 0.00618	G05F 0.0313	H04R 0.01583
D06F 0.00439	Y02D 0.0081	G07C 0.04747	A63H 0.00583	G01D 0.02347	G06V 0.01493
H04H 0.00374	G11B 0.0079	B25J 0.04634	H04S 0.00516	Y02P 0.02121	H04B 0.01037
F25D 0.00357	H04S 0.0059	G02B 0.04425	H01R 0.00415	G01R 0.01822	G06T 0.00843
Cluster 7	**Cluster 8**	**Cluster 9**	**Cluster 10**	**Cluster 11**	
G05B 0.56155	G08B 0.4918	G06Q 0.5009	G06N 0.0753	H04M 0.62993	
F24F 0.1505	G06V 0.08672	G05F 0.03718	G05D 0.07495	G08C 0.07596	
H04L 0.06813	H05B 0.0621	Y10S 0.03583	G07C 0.07014	H05B 0.03656	
G05D 0.04276	G01J 0.04156	Y04S 0.02804	G01S 0.06307	F24F 0.03047	
G05F 0.01886	G06T 0.03866	Y02P 0.02741	Y02A 0.06028	G06T 0.02794	
Y02P 0.01715	G01V 0.03786	G10L 0.02708	G01C 0.06024	G01V 0.02359	
G08C 0.01564	G01N 0.03641	G05D 0.02077	B60R 0.05902	G01N 0.02292	
G09B 0.00981	Y02A 0.02922	G06V 0.0206	G08G 0.05879	Y02A 0.02124	
H04Q 0.00851	F24F 0.01718	H05K 0.02042	B60W 0.05702	G01J 0.02111	
E06B 0.00788	E03B 0.01292	G16H 0.01528	B60L 0.05654	H04Q 0.01876	

By inspecting the above table, we are able to detect that there are several clusters that are mostly formed by a single CPC subclass. However, we can also observe that three clusters are formed by multiple subclasses, e.g., Cluster 10, meaning that our approach indeed discovered associations between different CPC encodings. It is also evident that while the prevalent subclasses correspond to the primary subclass for home automation, other subclasses that comprise the clusters relate to more distinct objectives. For example, Cluster 1 contains subclasses related to lightbulbs (F21V), refrigerators (F25D) and laundering (D06F), indicating that home automation networking expands across various domains. The different CPC subclasses, shown in Table 1, along with the descriptions of the representative CPC groups and subgroups of each cluster, finally help us provide a descriptive title for each cluster (Table 2). The reason behind the inspection and presentation of the most frequent CPC groups and subgroups of each cluster is because each patent is assigned to multiple identifiers of this level rather than to CPC subclasses directly.

Table 2. Cluster titles and prevalence.

Cluster Title	Representative CPC Groups and Subgroups	Overall Prevalence (Diagonal Values of d)
(1) Transmission of digital information	H04L12/2803 (2743 patents)	6650.377
(2) Electric digital data processing, e.g., interaction techniques based on graphical user interfaces	G06F3/0482 (304 patents) G06F3/167 (274 patents) G06F3/04847 (218 patents) G06F3/04842 (195 patents) G06F3/04817 (168 patents) G06F3/0484 (154 patents)	4497.858
(3) Transmission systems, i.e., transmission of electrical, optical, and radio signals, between computers and devices	G08C17/02 (444 patents) G08C2201/93 (161 patents) H04B3/54 (119 patents) H04B3/542 (93 patents) H04B2203/5445 (79 patents)	3760.012
(4) Wireless communication networks, e.g., specially adapted devices, communication services, discovery of network devices, access security	H04W4/80 (694 patents) H04W84/12 (410 patents) H04W4/70 (330 patents) H04W8/005 (241 patents) H04W12/08 (229 patents)	3567.315
(5) Efficient supplying and distributing of electric power, e.g., for mitigation of climate change	Y04S20/20 (294 patents) Y02B70/30 (332 patents) Y02B20/40 (242 patents) Y04S20/222 (131 patents) H02J2310/00 (119 patents)	3205.856
(6) Selective content distribution via interactive pictorial communication, e.g., interfacing home networks or client devices specifically adapted for the reception of or interaction with content	H04N21/43615 (845 patents) H04N21/43637 (251 patents) H04N21/41265 (233 patents) H04N21/42204 (219 patents) H04N21/482 (187 patents)	3044.548
(7) Domestic control or regulating systems, e.g., air conditioning	G05B15/02 (1155 patents) G05B2219/2642 (644 patents) F24F11/30 (387 patents) F24F11/58 (288 patents) F24F11/62 (249 patents)	2816.888
(8) Signaling or calling systems, e.g., alarm systems	G08B25/008 (163 patents) G08B25/08 (145 patents) G08B17/10 (131 patents) G08B25/10 (131 patents)	1705.421
(9) ICT, methods or systems specially adapted for administrative, commercial, financial, managerial or supervisory purposes	G06Q10/20 (141 patents) G06Q50/06 (140 patents) G06Q10/06 (119 patents)	1696.417
(10) Computing arrangements based on computational and learning models (especially in autonomous vehicles), e.g., temperature control, vehicle maintenance indicators, distance or positioning measurement, length or thickness measurement, and traffic control.	G06N20/00 (164 patents) G05D23/1917 (85 patents) G07C5/008 (110 patents)	1516.787
(11) Telephonic communication, e.g., remote control of appliances, combination with other systems/computers	H04M1/72415 (245 patents) H04M11/062 (86 patents)	1463.462

The most representative CPC subclasses and cluster titles indicate that the general clusters of home automation networking patents are related to both technical content, such as the transmission of digital information and electric digital data processing, and to relevant applications, such as alarm systems, air conditioning, and pictorial communication. In addition, we should mention that the first cluster is the most relevant to our dataset as the retrieved patent records belong to the CPC subgroups that fall under the H04L12/2803 CPC encoding. Of course, clusters with reduced prevalence present more interesting characteristics and refer to specific technologies and patent objectives such as home automation networking targeted at finance and management (Cluster 9), general-purpose alarm systems (Cluster 8), air conditioning and regulation systems (Cluster 7) or autonomous systems that utilize computation methods for multi-purpose usage in vehicles, e.g., temperature control or traffic monitoring (Cluster 10). Finally, an interesting indicator of the rise of IoT technologies is the dedicated cluster for remote appliance control and communication via sensors and other computer systems (Cluster 11).

Remarkably, by reviewing the patent abstracts of the retrieved patents, we noticed several common control units of smart home networks, including microcontrollers (Cluster 3, Cluster 7, Cluster 11), programmable logic controllers (Cluster 3, Cluster 5), tablets (Cluster 5, Cluster 7, Cluster 11) and phones/smartphones (Cluster 11). At the same time, we also observed multiple communication protocols such as Wi-Fi (Cluster 4), Ethernet (Cluster 3, Cluster 6, Cluster 11), ZigBee (Cluster 4, Cluster 5), Bluetooth (Cluster 2, Cluster 4, Cluster 9, Cluster 11), IEEE (Cluster 6, Cluster 9) and X10 (Cluster 1, Cluster 4). Our research shows that the different control units and communication protocols are not employed in similar technologies as they are not gathered around distinct clusters, meaning that they have advantages and disadvantages against each other in the various communication networks. To detect the aforementioned technologies, we traced patent abstracts that contained them and then analyzed the weight coefficients of each patent (matrix *h*) in order to denote the representative clusters that included these technologies.

4.2. RQ_2: *How Are the Different Technology Clusters Classified with Respect to Patent Activity Trends and Growth?*

According to Choi and Song [9], the four classes that are used to characterize each cluster are formed using two critical thresholds. The first threshold is the median cluster PS (x axis) while the second is the 0 CAGR (y axis). Thus, by using these two thresholds, we are able to create a two-dimensional space with 4 quartiles representing analogous classifications, as presented in Figure 5. More specifically, each quartile corresponds to combinations of PS and CAGR, with PS receiving only positive values and the CAGR receiving positive and negative values. Hence, clusters are represented as points in the two-dimensional space, falling into different quartiles which are interpreted as either dominant, emerging, declining or saturated. Evidently, dominant and emerging clusters are more important as they indicate technologies that are either the primary focus of patent objectives and assignees or have prospects for further investment. On the other hand, declining and saturated clusters can be traced to technologies that are rendered obsolete or that have been overshadowed by other technologies i.e., technologies that are present in dominant or emerging clusters.

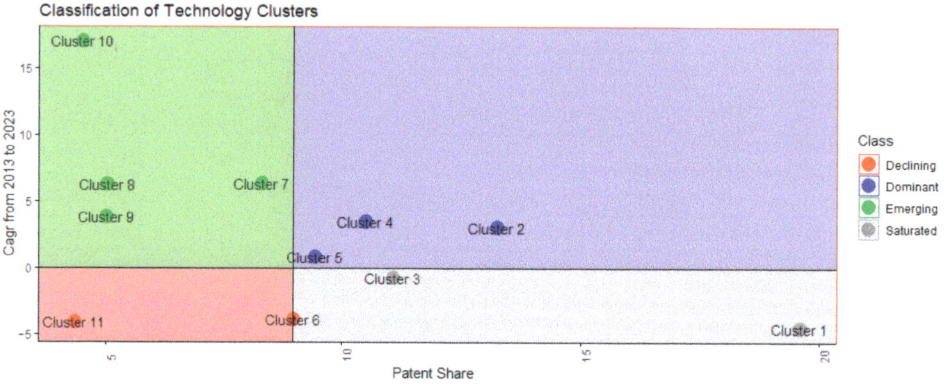

Figure 5. Classification of technology clusters.

Based on the aforementioned classifications, Figure 5 shows that Cluster 10 (computational systems for vehicle control) is the most emergent technology cluster (green quartile), having a relatively high CAGR and low PS. Hence, we can conclude that the recent trends of home automation networking technologies are related to computational models which are used for various purposes, with autonomous vehicles being the primary focus. This is a prime example of an emerging technology, as autonomous vehicles are indeed a field that has seen a rise in investments from high-profile companies and is in dire need of automated networking systems. Other clusters that represent emerging technologies are those that represent air conditioning and regulation systems (Cluster 7), multi-purpose alarm systems (Cluster 8) and automation technologies for financial and managerial services. Based on Table 2, these clusters, while having a low prevalence overall in our patent data, are not necessarily sectors that have no industrial interest but rather, they represent niche and domain-specific markets that are met with a rise in investments. In addition, Cluster 2 (electric digital data processing) should be considered as the most dominant technology cluster (blue quartile), meaning that a relatively high proportion of home automation networking patents process digital data and interface interactions, in turn meaning that there is also an increasing interest rate in these technologies. In general, in the dominant clusters category, we can discern technological clusters that, while not necessarily groundbreaking, represent standard practices in home automation networking that dominate the market due to the fact that most patents utilize them in their methodologies. This is further validated by the other dominant clusters, which are both essential parts of automation networking and refer to wireless communications (Cluster 4) and energy/power supply (Cluster 5).

At the same time, the declining clusters (red quartile) are mostly associated with telephonic and pictorial communications (Cluster 11). This observation shows that these two technological areas were trending prior to 2013, particularly during the rise of smartphones, but not in the more recent innovations, where other more innovative concepts have overtaken them. Finally, the saturated clusters (grey quartile) contain Cluster 3 (transmission systems) and Cluster 1 (digital information transmission). This is not a surprising fact, as traditional or analogue transmission systems can be considered a saturated field, one which, though an essential part of home automation networking, contains multiple solutions that perform different functions and have been complemented by developments in dominant and emerging fields (e.g., IoT, wireless communications). Additionally, as every patent of our dataset is assigned to the *H04L* subclass, the classification of Cluster 1 indicates that the recent home automation networking patents have an increased number of assigned technology areas, expressed by CPC subclasses and subgroups. A final finding is associated with the classification of Cluster 6 (interaction with content and pictorial communications) which balances between the declining and saturated categories, hence determining a technology field that, while not entirely uncompetitive, has been rendered moderately innovative and overshadowed by other relevant clusters (e.g., Cluster 2).

Overall, our analysis can be used as a knowledge basis in profiling the recent activity of home automation networking patents and assessing technology scouting directions for future inventions. For example, the distinguished emerging and dominant technology clusters should be considered as the primary options in forming a single home automation networking patent, focusing on prevalent sectors and technologies that provide a competitive advantage to companies that will invest in relevant patents. Similarly, the declining and saturated fields, while not necessarily obsolete, should be approached carefully when patenting products relevant to them.

4.3. RQ_3: Who Are the Leading Assignees/Competitors of the Different Technology Clusters of Home Automation Networking?

Our first assumption is that the most prevalent investors of the retrieved patents are expected to also be heavily involved in the different technology clusters. In Table 3, we present the 10 most involved assignees in the retrieved home automation networking patents. The main activities of these assignees are related to the manufacturing of electronic equipment, e.g., software and hardware for smartphones, televisions, and the development of semiconductors. Among these, we can see well known technological companies that have established footholds in technological areas such as hardware and electronics (AT&T, Samsung, Panasonic, LG), semiconductors (Broadcom Corporation) and software or computing (Google, Intel). This is another indicator that home automation networking is a multidisciplinary field that attracts the interest of high-profile investors, who strive to patent their products and exploit them commercially.

Table 3. The top 10 assignees of home automation networking patents.

Assignee	No Patents
Samsung Electronics Co., Ltd.	582
SONY Corporation	371
Google LLC	316
LG Electronics Inc.	250
AT&T Intellectual Property I, L.P.	182
Broadcom Corporation	175
SONY Electronics Inc.	147
Sonos, Inc.	139
Intel Corporation	110
Panasonic Intellectual Property Corporation of America	108

Furthermore, the established AC matrix shows that some technology clusters are mostly associated with a single assignee, while others are related to multiple. For example, the weight coefficients of the patents (matrix *h*) indicate that Cluster 10 (Computational systems for Vehicle control) is formed by patents that are mostly owned by the State Farm Mutual Automobile Insurance Company (40%). At the same time, similar evaluations are inspected for Cluster 3, which is related to transmission systems, (Intel Corporation—18%) and Cluster 8, which is related to alarm systems, (Google LLC—21%). The remaining clusters are distributed to several assignees in a more balanced way, but we can also distinguish some that are more prevalent. In Figure 6, we present the most prevalent assignees of each technology cluster with respect to the classification of the clusters (see Figure 5).

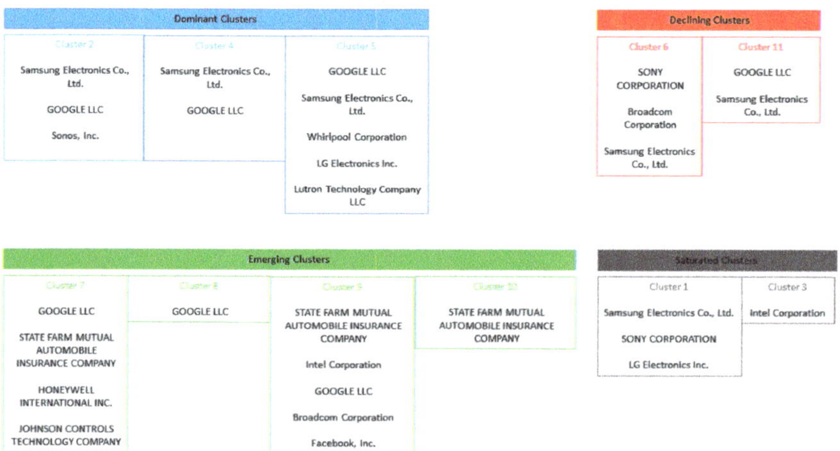

Figure 6. Most prevalent assignees per technology cluster.

The results from our analysis show that the primary competitors of the different technology clusters are indeed the major investors of the whole patent dataset. In particular, we observe that Google LLC and Samsung Electronics Co., Ltd. are among the main competitors in seven and six technology clusters, respectively. Likewise, these two assignees hold a significant stake in the dominant technology clusters, while Google LLC is also apparent in three out of the four emerging clusters as well. This proves that, where dominant and emerging technologies of home automation networking are concerned, software solutions and innovative hardware play a crucial role and are considered the top choices for patenting and protecting intellectual property.

Furthermore, we should emphasize that State Farm Mutual Automobile Insurance Company, despite its overall low involvement (89 patents) when compared with the major investors, holds a large proportion of patents that are relevant to three out of the four emerging clusters. The findings indicate that this company brought to the surface some new technologies that are related to home automation networking, as they own patents with a cumulative weight coefficient that exceeds forty percent in the most emergent cluster, Cluster 10. In addition, the fact that this assignee is present in the emerging clusters category is an encouraging sign of a company that provides innovative solutions, not only for autonomous vehicles but also for regulation systems, pictorial communications and data management. The declining and saturated clusters also present some interesting findings, with Broadcom Corporation being present in one declining cluster, related to content interaction (Cluster 6), while having a notable presence in granted patents, according to Table 3. In addition, Sony, while being one of the top ten assignees, is also present only in the declining and saturated clusters. This does not necessarily mean that these companies do not have robust patenting strategies but rather that the dominant and emerging categories are populated with companies more akin to software development and hardware. Finally, in the dominant clusters, several known hardware companies are present (Whirlpool Corporation, LG Electronics Inc.) while the presence of Sonos Inc., a company that sells sound-related products, reveals interesting technological avenues of dominant patents.

Therefore, we believe that researchers and companies should study the technologies, strategies and business models of the companies that are present in the dominant and emerging clusters, in order to create new or to expand existing clusters and direct their patenting objectives towards these technologies, due to their contribution and general involvement in home automation networking patents.

In summary, in this subsection we have provided information concerning a primary task related to patent analysis—competitor analysis—as we identified the main competitors of each technology cluster. Therefore, inventors and organizations may leverage our outcomes to assess their immediate competitors and to further gain insights into their strategies, tactics and products. Through this process, an organization can detect misplaced or misdirected actions in its strategies while also pinpointing some potential technological gaps that would offer a significant advantage against these competitors.

5. Discussion

Overall, in this study we have explored the main technologies which are relevant and which contribute to home automation networking inventions. Through our analysis, we have shown that the different technology clusters encompass various areas of interest that are relevant to both technical content, e.g., transmission systems and data processing, and to specific appliances, e.g., alarm systems, air conditioning or financial systems (RQ_1).

Moreover, we have classified the main technology clusters by evaluating their overall development across the past decade (2013–2023). In short, the respective outcomes have revealed that home automation networking patents are associated with an increased number of relevant technologies in this period. The detailed mapping distinguishes computational models and data processing techniques as the most emergent and dominant clusters, respectively, along with several upcoming concepts in home automation networking, such as wireless technologies and more efficient energy/power supply methodologies (RQ_2).

In addition, the extracted results indicate that the retrieved patent data also contain several essential hardware technologies and protocols that are being leveraged in home automation networking. Among these, we can observe equipment such as tablets and PLCs and protocols such as Wi-Fi or Ethernet and Bluetooth communication. The existence of these systems in patent data proves that home automation networking concerns multiple interested parties, including software, hardware and firmware companies.

Finally, the available patent data have helped us to assess the leading assignees of each technology cluster using the portfolio of each assignee and the scores/coefficients linking each patent with the different clusters. The results point out some major assignees that invest in many dominant technology areas, e.g., Google LLC and Samsung Electronics Co., Ltd., while also some assignees that are more centralized to home automation networking appliances, e.g., State Farm Mutual Automobile Insurance Company. By leveraging the available cluster classifications, we conclude that the latter assignee has a strong presence in the most emerging technologies while the former two assignees should be characterized as the most dominant. In addition, some well-known hardware companies, such as Whirlpool and LG, have found secure pathways in patenting and have invested in dominant and emerging technological clusters, while Broadcom Corporation and Sony were only detected in the declining and saturated clusters. These findings have helped us ensure that RQ_3 is addressed in detail.

Regarding the practical implications of this study, we believe that the employed framework and the findings from our experiments contribute to the provision of insights to both researchers and industrial actors.

In terms of research interest and implications, we have introduced an approach by which to assess technology clusters from patent data using patent classifications and NMF, which is indeed an algorithm which has been previously and effectively utilized for multiple tasks and data types. Compared with other clustering techniques, our approach assesses the links between each patent and each cluster with scoring mechanisms, instead of assigning a single cluster to each patent. As each patent is assigned to multiple CPC subgroups, and by extension to several subclasses, this practice seems a more appropriate approach than the standard hard clustering techniques. Additionally, the results and methodological framework of this study can be easily reproduced following our detailed approach and using various implementations of NMF that are publicly available on multiple programming languages, e.g., Python, R. Thus, the employed framework can be adapted to other datasets for similar patent analysis tasks or altered according to the objectives of each researcher. As many patent offices make use of the CPC or the International Patent Classification (IPC) schemas, which are similar in terms of structure and identifiers, our framework can be adapted to other countries and datasets beyond those belonging to the USPTO. Of course, for such an analysis to be carried out effectively, the structures and potentially varied patenting and legal procedures of other offices should be taken under careful consideration. In addition, apart from patent data, our approach could be useful in regard to different data types that use classification schemas or tagging systems, e.g., research papers, Twitter posts, Q&A discussions.

Furthermore, from a business perspective, the outcomes of this study provide insights concerning the main technologies and assignees of home automation networking patents. Among the different use cases, our results can be used as a basis when establishing high quality and valuable inventions, by providing information about emerging and declining technologies. Additionally, as we evaluated the involvement of the assignees in the different technology clusters, both researchers and practitioners or industrial actors can study our outcomes to identify immediate and major competitors to further build business strategies or to accordingly study groups of assignees. Finally, individuals or organizations which aim to establish or study home automation networks can leverage our findings and detect the general and more niche or refined technologies which can be used to establish relevant systems and networks, as the initial CPC identifiers are numerous compared with the eleven technology clusters which were assessed by our approach.

6. Threats to Validity

Although we based our framework on existing approaches and validated data sources, we identified several threats that can be categorized into the general concepts of internal and external validity.

In our case, the internal threats concern the employed methodologies, data structures and assessment methods. First, by exclusively using the CPC subclasses as our knowledge basis, we may have omitted the valuable information included in the descriptions and titles of the retrieved patents and the general intricacies of semantic data that may also contain technological objectives. However, existing studies have also exclusively studied patent classifications in order to answer research objectives that are similar to the RQs of our study. In addition, the CPC identifiers provide detailed information on the specifications of each patent as they are assigned by the experts of the related fields and the applicants themselves. Hence, we deem the threat of information loss to be mitigated by the fact that we utilized a well-known classification schema that is assigned after careful expert judgement.

Moreover, the clustering approach that was followed is currently a less explored one, as NMF is not commonly used in patent analysis. This fact raises the potential of bias in the selection of the algorithms, which may also affect the validity of the respective outcomes. Nevertheless, NMF is a standard approach to the discovery of underlying patterns from various data types and is also validated as an effective approach in different tasks. Finally, the assessment methods that were used to find the optimal rank, i.e., number of technology clusters, the classes of the different clusters, and the prevalence of each assignee, also fall under this potential for bias. To overcome issues of this nature, we based our analysis on existing techniques and assessment methods, e.g., NPMI, PS, and CAGR, that are commonly employed in both similar and in more generic tasks.

The external validity refers to the significance and potential generalization of the study's outcomes. First, we believe that the choice of including a single patent office in our analysis could raise some issues concerning the generalization of our findings. Despite these vital concerns, the USPTO is considered the primary choice for patent analysis research, capturing a more global perspective than other patent offices [39], while also using a respected classification schema for the storage of information. Furthermore, additional data sources, including business reports and related studies, could help us form a more inclusive/generic dataset that could also capture the internal and external factors that formulate technological trends and are not directly relevant to patent grants. However, patent data indicate the industrial activity of the different firms in a concise and contained way and thus the employed dataset can be considered a reliable knowledge base.

7. Conclusions

In this section, we present the main conclusions of this study as indicated by the employed framework and the respective findings. Our analysis has shown that the employed framework is able to extract coherent and distinct technology clusters that were easily interpretable. In addition, it has also addressed and uncovered the associations between different technologies which combine to create a home automation networking invention. The outcomes of our approach have shown that this field is characterized by a variety of different technologies and applications which aim to enhance automated procedures, which in turn ensure safety, comfort and control.

Moreover, by classifying the different clusters into four distinct categories of technological growth, we have provided information on the technologies that may be either reduced or replaced in the future, i.e., telephonic communication technologies. At the same time, we have distinguished both dominant and emerging technologies, i.e., data processing techniques and computational models, respectively, further providing guidelines for future directions. Lastly, we have assessed the leading assignees of the different technology clusters, hence revealing some popular assignees that invest in many different fields, e.g., Google LLC, and also some more domain-specific investors that obtained a lower involvement but a significant stake in emerging technologies, e.g., State Farm Mutual Automobile Insurance Company.

In summary, we believe that this study has produced valuable insights on home automation networking inventions, satisfying some standard patent analysis objectives such as technology forecasting, trending analysis and competitor analysis. We have further proposed an approach in the identification of technology clusters from patent data, which can also serve as a reproducible and valuable tool in future research.

8. Future Work

In this section, we discuss, in brief, some research directions for future work that could stand as extensions to this study or other existing works. First, we believe that additional prospects, such as patent descriptions, and data sources, such as literature data, may also provide significant information and extend or validate the findings of this study. Thus, researchers could investigate some publicly available and relevant data sources. In addition, beyond data clustering, there are other approaches that have been previously proposed and employed in prior work for research objectives similar to our own. Overall, we recommend co-word analysis and topic modelling as two effective alternatives in analyzing the main content of patent data.

Apart from the properties that describe the nature of the patents and their respective assignees, the inventors and the citation information have been previously studied as important patent characteristics. By analyzing the information of the different inventors of a patent dataset, researchers can identify the most productive and prolific as well as establish collaboration networks between their countries. Furthermore, the citation characteristics are usually studied as indicators that correspond to patent value and influence. Hence, researchers could leverage this information and further assess influential clusters by investigating patent citation networks or combine this information within the framework that we employed. This would lead to a more comprehensive and detailed analysis and forecasting of technologies while also taking into account geographical, interpersonal and research related trends.

Author Contributions: Conceptualization, K.C. and K.G.; data curation, K.C. and K.G.; formal analysis, K.G. and N.M.; investigation, K.C. and L.A.; methodology, K.C., K.G. and L.A.; project administration, L.A.; resources, K.C. and N.M.; software, K.C. and N.M.; supervision, N.M. and L.A.; validation, K.C., K.G. and N.M.; visualization, K.C. and K.G.; writing—original draft, K.C.; writing—review and editing, K.C., K.G., N.M. and L.A. All authors have read and agreed to the published version of the manuscript.

Funding: This research received no external funding.

Data Availability Statement: Data are available in a publicly accessible repository: https://github.com/koncharman/patents_smart_home (accessed on 19 September 2023).

Acknowledgments: The research in this paper is part of the Ph.D. dissertation of the first author.

Conflicts of Interest: The authors declare no conflict of interest.

References

1. Kadam, R.; Mahamuni, P.; Parikh, Y. Home automation system. *Int. J. Innov. Res. Adv. Eng.* **2015**, *2*, 81–86.
2. Sovacool, B.K.; Del Rio, D.D.F. Home automation technologies in Europe: A critical review of concepts, benefits, risks and policies. *Renew. Sustain. Energy Rev.* **2020**, *120*, 109663. [CrossRef]
3. Marikyan, D.; Papagiannidis, S.; Alamanos, E. A systematic review of the home automation literature: A user perspective. *Technol. Forecast. Soc. Chang.* **2019**, *138*, 139–154. [CrossRef]
4. Zielonka, A.; Woźniak, M.; Garg, S.; Kaddoum, G.; Piran, M.J.; Muhammad, G. Home automations: How much will they support us? A research on recent trends and advances. *IEEE Access* **2021**, *9*, 26388–26419. [CrossRef]
5. Guderian, C.C. Identifying emerging technologies with smart patent indicators: The example of smart houses. *Int. J. Innov. Technol. Manag.* **2019**, *16*, 1950040. [CrossRef]
6. Abbas, A.; Zhang, L.; Khan, S.U. A literature review on the state-of-the-art in patent analysis. *World Pat. Inf.* **2014**, *37*, 3–13. [CrossRef]

7. Kim, G.; Bae, J. A novel approach to forecast promising technology through patent analysis. *Technol. Forecast. Soc. Chang.* **2017**, *117*, 228–237. [CrossRef]
8. Altuntas, S.; Erdogan, Z.; Dereli, T. A clustering-based approach for the evaluation of candidate emerging technologies. *Scientometrics* **2020**, *124*, 1157–1177. [CrossRef]
9. Choi, D.; Song, B. Exploring technological trends in logistics: Topic modeling-based patent analysis. *Sustainability* **2018**, *10*, 2810. [CrossRef]
10. Lee, D.; Seung, H.S. Learning the parts of objects by non-negative matrix factorization. *Nature* **1999**, *401*, 788–791. [CrossRef]
11. Lee, D.; Seung, H.S. Algorithms for non-negative matrix factorization. In *Advances in Neural Information Processing Systems 13*; The MIT Press: Cambridge, MA, USA, 2000.
12. Shuhaiber, A.; Mashal, I. Understanding users' acceptance of home automations. *Technol. Soc.* **2019**, *58*, 101110. [CrossRef]
13. Shin, J.; Park, Y.; Lee, D. Who will be home automation users? An analysis of adoption and diffusion of home automations. *Technol. Forecast. Soc. Chang.* **2018**, *134*, 246–253. [CrossRef]
14. Toschi, G.M.; Campos, L.B.; Cugnasca, C.E. Home automation networks: A survey. *Comput. Stand. Interfaces* **2017**, *50*, 42–54. [CrossRef]
15. Yu, J.; de Antonio, A.; Villalba-Mora, E. Deep learning (CNN, RNN) applications for home automations: A systematic review. *Computers* **2022**, *11*, 26. [CrossRef]
16. Ma, S.C.; Xu, J.H.; Fan, Y. Characteristics and key trends of global electric vehicle technology development: A multi-method patent analysis. *J. Clean. Prod.* **2022**, *338*, 130502. [CrossRef]
17. Giczy, A.V.; Pairolero, N.A.; Toole, A.A. Identifying artificial intelligence (AI) invention: A novel AI patent dataset. *J. Technol. Transf.* **2022**, *47*, 476–505. [CrossRef]
18. Charmanas, K.; Mittas, N.; Angelis, L. Topic and influence analysis on technological patents related to security vulnerabilities. *Comput. Secur.* **2023**, *128*, 103128. [CrossRef]
19. Georgiou, K.; Mittas, N.; Ampatzoglou, A.; Chatzigeorgiou, A.; Angelis, L. Data-Oriented Software Development: The Industrial Landscape through Patent Analysis. *Information* **2022**, *14*, 4. [CrossRef]
20. Jee, J.; Shin, H.; Kim, C.; Lee, S. Six different approaches to defining and identifying promising technology through patent analysis. *Technol. Anal. Strateg. Manag.* **2022**, *34*, 961–973. [CrossRef]
21. Park, M.; Geum, Y. Two-stage technology opportunity discovery for firm-level decision making: GCN-based link-prediction approach. *Technol. Forecast. Soc. Chang.* **2022**, *183*, 121934. [CrossRef]
22. Geum, Y.; Kim, M. How to identify promising chances for technological innovation: Keygraph-based patent analysis. *Adv. Eng. Inform.* **2020**, *46*, 101155. [CrossRef]
23. Suominen, A.; Toivanen, H.; Seppänen, M. Firms' knowledge profiles: Mapping patent data with unsupervised learning. *Technol. Forecast. Soc. Chang.* **2017**, *115*, 131–142. [CrossRef]
24. Daim, T.; Lai, K.K.; Yalcin, H.; Alsoubie, F.; Kumar, V. Forecasting technological positioning through technology knowledge redundancy: Patent citation analysis of IoT, cybersecurity, and Blockchain. *Technol. Forecast. Soc. Chang.* **2020**, *161*, 120329. [CrossRef]
25. Wang, J.; Hsu, C.C. A topic-based patent analytics approach for exploring technological trends in smart manufacturing. *J. Manuf. Technol. Manag.* **2021**, *32*, 110–135. [CrossRef]
26. Charmanasa, K.; Georgioua, K.; Mittasb, N.; Angelisa, L. Unveiling technology clusters and prominent investors of home automation networking through patent analysis. *Algorithms* **2023**, *9*, 23.
27. De Bruine, Z.J.; Melcher, K.; Triche, T.J. High-performance non-negative matrix factorization for large single-cell data. *bioRxiv* **2021**. [CrossRef]
28. Wang, Y.X.; Zhang, Y.J. Nonnegative matrix factorization: A comprehensive review. *IEEE Trans. Knowl. Data Eng.* **2012**, *25*, 1336–1353. [CrossRef]
29. Lee, S.; Kim, M.; Shin, S.; Park, S.; Jeong, Y. Data-dependent feature extraction method based on non-negative matrix factorization for weakly supervised domestic sound event detection. *Appl. Sci.* **2021**, *11*, 1040. [CrossRef]
30. Shimada, K.; Bando, Y.; Mimura, M.; Itoyama, K.; Yoshii, K.; Kawahara, T. Unsupervised speech enhancement based on multichannel NMF-informed beamforming for noise-robust automatic speech recognition. *IEEE/ACM Trans. Audio Speech Lang. Process.* **2019**, *27*, 960–971. [CrossRef]
31. Laxmi Lydia, E.; Krishna Kumar, P.; Shankar, K.; Lakshmanaprabu, S.K.; Vidhyavathi, R.M.; Maseleno, A. Charismatic document clustering through novel K-Means non-negative matrix factorization (KNMF) algorithm using key phrase extraction. *Int. J. Parallel Program.* **2020**, *48*, 496–514. [CrossRef]
32. Hassani, A.; Iranmanesh, A.; Mansouri, N. Text mining using nonnegative matrix factorization and latent semantic analysis. *Neural Comput. Appl.* **2021**, *33*, 13745–13766. [CrossRef]
33. Rajendra Prasad, K.; Mohammed, M.; Noorullah, R.M. Visual topic models for healthcare data clustering. *Evol. Intell.* **2021**, *14*, 545–562. [CrossRef]
34. Navas-Auger, W.; Manian, V. Spatial Low-Rank Tensor Factorization and Unmixing of Hyperspectral Images. *Computers* **2021**, *10*, 78. [CrossRef]
35. Peng, S.; Ser, W.; Chen, B.; Lin, Z. Robust semi-supervised nonnegative matrix factorization for image clustering. *Pattern Recognit.* **2021**, *111*, 107683. [CrossRef]

36. Bhandari, N.; Pahwa, P. Achieving Data Privacy Using Extended NMF. In *Machine Intelligence and Data Science Applications: Proceedings of MIDAS*; Springer Nature: Singapore, 2021; pp. 211–225.
37. Luo, X.; Liu, Z.; Jin, L.; Zhou, Y.; Zhou, M. Symmetric nonnegative matrix factorization-based community detection models and their convergence analysis. *IEEE Trans. Neural Netw. Learn. Syst.* **2021**, *33*, 1203–1215. [CrossRef]
38. Bouma, G. Normalized (pointwise) mutual information in collocation extraction. *Proc. GSCL* **2009**, *30*, 31–40.
39. Kim, J.; Lee, S. Patent databases for innovation studies: A comparative analysis of USPTO, EPO, JPO and KIPO. *Technol. Forecast. Soc. Chang.* **2015**, *92*, 332–345. [CrossRef]

computers

Article

QoS-Aware and Energy Data Management in Industrial IoT

Yarob Abdullah and Zeinab Movahedi *

School of Computer Engineering, University of Science and Technology of Iran, Tehran 1684613114, Iran; yarob.abdullah@comp.iust.ac.ir
* Correspondence: zmovahedi@iust.ac.ir

Abstract: Two crucial challenges in Industry 4.0 involve maintaining critical latency requirements for data access and ensuring efficient power consumption by field devices. Traditional centralized industrial networks that provide rudimentary data distribution capabilities may not be able to meet such stringent requirements. These requirements cannot be met later due to connection or node failures or extreme performance decadence. To address this problem, this paper focuses on resource-constrained networks of Internet of Things (IoT) systems, exploiting the presence of several more powerful nodes acting as distributed local data storage proxies for every IoT set. To increase the battery lifetime of the network, a number of nodes that are not included in data transmission or data storage are turned off. In this paper, we investigate the issue of maximizing network lifetime, and consider the restrictions on data access latency. For this purpose, data are cached distributively in proxy nodes, leading to a reduction in energy consumption and ultimately maximizing network lifetime. To address this problem, we introduce an energy-aware data management method (EDMM); with the goal of extending network lifetime, select IoT nodes are designated to save data distributively. Our proposed approach (1) makes sure that data access latency is underneath a specified threshold and (2) performs well with respect to network lifetime compared to an offline centralized heuristic algorithm.

Keywords: data access latency; energy-aware data management; Industry 4.0; IoT; maximizing network lifetime; proxy node

Citation: Abdullah, Y.; Movahedi, Z. QoS-Aware and Energy Data Management in Industrial IoT. *Computers* **2023**, *12*, 203. https://doi.org/10.3390/computers12100203

Academic Editor: Sergio Correia

Received: 19 August 2023
Revised: 19 September 2023
Accepted: 22 September 2023
Published: 10 October 2023

1. Introduction

One of the most important improvements in the recent technological universe is the IoT. The IoT involves connecting and integrating billions of smart devices and networks, such as wireless sensor networks (WSNs), to the internet. This creates networks that can share and interchange data to increase performance and, ultimately, individual interaction. IoT applications span a wide range of fields, including transportation, smart building control, energy management through smart meters, healthcare services, and home automation [1].

Industrial automation is currently undergoing a significant transformation, thanks to the advent of IoT technology in industrial applications. This transformation has become possible due to recent technological advancements that enable extensive and precise interconnectivity. Efforts to automate processes independently of continuous human intervention rely on the seamless flow of data between sensors, controllers, and actuators on a large scale. In recent times, the focus has been on developing and optimizing data interchange and distribution schemes within industrial structures. Data generated in this context are typically transmitted wirelessly to a central network controller. The controller then analyzes the received data and, when necessary, adjusts network pathways and data transfer mechanisms. This process not only optimizes resource allocation but also influences physical environments through actuator systems.

In industrial networks, topologies and connectivity can vary due to connection or sensor node defeats. Additionally, highly dynamic situations, where connection efficiency differs significantly from central scheduling calculations, may result in sub-optimal efficiency and possibly lead to the construction of non-guaranteed application needs. These dynamic network topologies can cause several nodes of industrial sensors to fail. The increase in systems that have batteries causes industrial networks to consume a lot of energy. Taking advantage of locally distributed computation exceeds what would normally be required [2].

In order to meet critical requirements, such as latency and network lifetime in industrial applications, data management should be based on a flexible and reliable architecture. Generating a large number of data was rarely investigated in the past and was less considered due to the problems that existed in the analysis of large volumes of data. But today, by using data management methods, this important issue can be addressed, and valuable information can be obtained in this field. It should be noted that for data management, data characteristics are investigated based on practical cases.

A large number of sensor nodes use batteries. Therefore, limiting the amount of energy in each node is one of the important challenges of industrial networks. One straightforward procedure involves delivering the packet to its destination with minimal power consumption. One popular solution involves using the shortest path with a connection cost that is the same as the energy required on every link to transmit the packet. Another method involves the maximization of network lifetime [3]. The definition of the lifetime of a WSN can vary, but it is often expressed as the elapsed time from when the first node loses power. The intended meaning of the sentence is to convey that the lifetime of a WSN can be defined in various ways. One possible definition is based on the time elapsed from the beginning of data distribution to the moment the first node in the network depletes its energy.

The model that is very common for data transfer in the industrial IoT is pub/sub [4]. The implementation of this model in industrial IoT may not be applicable due to high energy consumption and data access delay. To adopt the pub/sub industrial IoT, several papers are available that illustrate distributed methods. To explore the implementation of pub/sub mechanisms within industrial IoT contexts, numerous papers provide insights into distributed methodologies. Notable examples include [2,5,6], which collectively examine the utilization of specialized, high-capacity nodes for data storage. In these studies, the focus lies on employing select nodes with enhanced capabilities, setting them apart from standard nodes, to effectively manage and store data. In these works, several nodes that are more powerful and different from other nodes have been used to store data. Despite the outstanding works in the mentioned model, there are numerous areas ripe for development and progress; the research conducted is still in the early stages

The growth of IoT devices is leading to massive amounts of data that require low-latency access and processing from cloud data centers. This drives the need for efficient resource management and network optimization [7,8]. Battery-powered IoT devices, like sensors, have limited energy; thus, methods to reduce power consumption through scheduling, duty-cycling, and energy-harvesting are important [8,9]. For networks with many battery-constrained devices, like sensors, the lifetime is critical and can be extended through efficient protocols, scheduling, duty-cycling, and energy-harvesting techniques [7,9]. Energy-efficient distributed mobile data management is a promising approach that uses local proxies and network optimization to provide low-latency access while also saving energy [7–9].

This paper addresses strict latency requirements and introduces an energy-aware data management method (EDMM) that maximizes network lifetime designed to distribute and cache data at selected proxy nodes near sensors and actuators. This significantly reduces energy consumption, latency, and overhead, aligning with the principles of environmentally friendly industrial IoT practices. Dynamic node management strategies are incorporated, ensuring that nodes that are not actively involved in caching or communication are switched off to conserve energy resources. These innovations collectively pave the way for more responsive and efficient data management techniques in industrial IoT networks, aligning with the real-time demands of industrial applications.

In this article, the industrial IoT system, which consists of sensor and operator nodes, is considered. In the proposed model, data consumers are introduced as actuators and producers as sensors. Some intermediate nodes, which have different capabilities from other nodes, act as proxy nodes. The primary objective of this paper is to maximize the network's lifetime, taking into account certain limitations to enhance the performance of the proposed strategy. Given the location of the proxy, the initially limited energy resources, the data request models, and the maximum latency, this goal is achieved. For better performance, the nodes that are not involved in the process are turned off. Data are also prioritized and some data are available faster than others; these data are known as urgent data. In this way, to check the latency requirements, the data are considered in two categories—urgent and normal data—and each node has its own latency threshold. We show that the proposed method (1) guarantees data access latency below a specified threshold and (2) performs well in terms of network lifetime when compared to an offline centralized heuristic algorithm.

The remainder of this article is as follows: We supply a summary overview of the literature review in Section 2. The introduction of the model system is presented in Section 3. We illustrate the proposed approach (EDMM) in Section 4. In Section 5, the performance evaluation and obtained results are mentioned. Finally, in Section 6, the article is concluded, and we present some intuitions for future schemes.

2. Literature Review

Related works for the current study are [5,10,11]. In these articles, the authors focus on how to place the proxy in the network. In these works, the delay of data access is investigated and analyzed in order to improve the efficiency of the suggested approaches. Maximizing network lifetime is considered in [5]. A limited number of edge nodes, which have distinct and more powerful abilities than other nodes, are introduced as proxy nodes. Among the other objectives in that article, the location of proxies, the limited energy resources that nodes have, and the maximum delay that can be tolerated by consumer nodes are looked at. To maximize the lifetime of the network, the authors prove that the investigated problem is NP-hard and should be investigated by heuristic algorithms. In their proposed method, the authors show that the access latency is lower than the threshold and even though the lifetime of the network in their proposed method is lower than the optimal method, the performance of this method is better. The authors in [5] only consider the paths that achieve the maximum delay limit; all paths are not considered in this approach and the number of proxy nodes in the network is fixed.

In [10], the authors focus on energy consumption optimization. They consider the access latency, cache valency, and different data types in their investigation. In their proposed approach, energy consumption is considered in two ways, i.e., from the sensor to the proxy and from the proxy to the sensor. Regarding energy savings, some proxy nodes that are not involved in the process are turned off. In addition to considering proxies in the off mode, some data that are available more rapidly than others are designated as critical (urgent) data, and other data are designated as normal data. According to the data classification, the limits related to normal and critical data are separately considered; the limitation discussed in this article pertains to combined aspects of access latency. Moreover, along with the approach proposed by the authors, an algorithm based on ACS is also presented, and the meta-heuristic algorithm works like the optimal method in many cases. It should be noted that the proposed method is proven superior against corresponding methods based on different criteria, including energy consumption, access latency, and computing time. In [10], the authors do not consider the issue of maximizing the lifetime of the network and only focus on the optimization of energy consumption.

In [11], the authors consider a function that consists of the amount of energy consumption and access delay. This function includes the total energy consumption required to keep the nodes active and the energy consumption for data transfer, involving transmissions from the sensor to the proxy and from the proxy to the actuator. This function also includes data access latency. Their proposed method ensures the mean access latency remains below a predetermined maximum threshold, corresponding to data volume. The proposed strategy in [11] is similar to the strategy in [10], i.e., when increasing the efficiency of the proposed approach, data that are available faster are introduced as urgent data, and others are introduced as normal data. In this article, several proxy nodes are considered as idle (off). Finally, the proposed approach in 3 exhibits superior performance compared to similar and corresponding methods. The authors of [11] do not address the issue of maximizing network lifetime; their main goal is to reduce energy consumption.

In [12], the authors proposed new optimization formulas to maximize the network lifetime. Based on column generation, a method is provided to solve this type of optimization problem. In this article, the machine-to-machine connection is considered. In a machine-to-machine connection, sensor measurements are conducted within the network and are dispatched to various destinations through multi-part transmission. Since only a few configurations are used to maximize network lifetime, their proposed method is effective in practice. In addition to maximizing network lifetime, the authors provide upper and lower bands for their proposed formula. The authors do not address the issue of data access delay and do not pay attention to the role of proxy in the network.

The authors of [9] propose an energy-efficient resource management framework for software-defined data centers (SDDCs) to handle rapidly growing IoT and big data workloads. The consolidated model optimizes VM deployment and network bandwidth allocation to minimize energy consumption in SDDCs while guaranteeing quality of service. It uses a priority-aware heuristic approach based on weighted utility functions to select the best hosts and switches for allocating VMs and bandwidth for both critical and non-critical applications. The utility functions account for power consumption, resource utilization, and bandwidth usage. Experiments demonstrate that compared to existing schemes, the framework reduces the total energy consumption of SDDCs by 27.9%, with negligible quality of service violations of 0.33. The scheme is shown to be effective at improving energy efficiency in cloud data center resource management.

The authors of [8] propose an Internet of Things-based industrial data management framework with five layers: physical, network, middleware, database, and an application to efficiently collect and leverage massive, heterogeneous manufacturing data from smart devices on factory floors. The middleware layer collects, pre-processes, and aggregates real-time data using protocols like OPC-UA and provides modules for resources, events, data, and recovery management. A distributed database layer offers local storage prior to cloud transmission to avoid network delays. The application layer analyzes the data to gain insights into optimizing manufacturing processes, predicting maintenance, and driving smart factory decision-making. A case study with smart pumps demonstrates the framework's ability to successfully acquire, manage, and convert real-time industrial big data at scale into useful information to improve factory operations and productivity.

The authors of [7] explore technological trends that drive the evolution of massive MIMO into the 6G era, including metasurface-enabled massive MIMO for enhanced beamforming and sensing, ultra-massive MIMO at THz frequencies (offering tremendous capacity along with design challenges), cell-free architectures to improve spectral and energy efficiency, the integration of AI for gains in resource allocation and channel estimation, adaptations like non-coherent demodulation for high-speed applications, and expanding the reach to non-terrestrial networks while managing large delays and losses. The survey examines how these advancements, including intelligent surfaces, new frequency bands, innovative architectures, AI, and expanding applications, are transforming massive MIMO capabilities to meet future demands, but also require solutions to new challenges around factors like beam management, interference, transport, and modeling, to fully unlock their potential in 6G and beyond.

In [6], the authors specify and select a limited set of proxy nodes to store the data required by the consumer nodes, striking a balance between threshold data access latency and choosing a low number of proxies. The selection of proxy nodes should ensure guaranteed maximum access latency for data delivery to the requesting nodes. Any node can potentially be selected as a proxy node, and if the selection of proxy nodes is conducted correctly, the authors' goal of reducing access latency will be achieved. By minimizing the number of proxy nodes, the overall consumption of system resources is reduced. In this method, the average access latency is considered instead of the access latency of each node, and the maximization of network lifetime is not considered.

Standard WirelessHART uses graph routing to improve network reliability. The issue of network lifetime in graph routing is an important topic and has been focused on by many authors. The maximum lifetime of network WirelessHART under graph routing is mentioned in [13]; the authors prove that this problem is NP-hard and should be solved with the help of optimization algorithms. Therefore, in order to maximize the lifetime of WirelessHART networks, they introduce several algorithms. They show that the computation time required by greedy heuristics is greatly reduced, especially for WirelessHART networks, where graph roots may be computed often when network variations occur in open environments; thus, it is suitable and has good performance.

In [14], the authors develop their work from [5]. Considering the access delay, they attempt to increase network lifetime in industrial environments that have several hops. They prove that the problem is computationally complex and unsolvable; in order to solve the objective function, they design a one-step algorithm. Here, the authors use a fixed number consisting of proxy nodes and do not consider other modes of the proxy selection, such as whether the proxy nodes are on or off.

Sensor nodes in the WSN are nodes that have lower costs and less capability. However, they have the ability to work in environments that cannot be closed but cannot be transported in an effective manner. In [15], the authors propose a clustering technique to partition these nodes. In the clustering method, the cluster head must have special privileges, and the cluster heads are responsible for sending information to other nodes. In [15], the authors present a model for choosing the cluster head; the chosen method aims to maximize the lifetime of the network and optimize energy consumption. This method takes into account limitations, such as lower energy consumption and delay. The authors compare their proposed method with different algorithms and prove that this method exhibits superior performance. To achieve the article's goal, the authors utilize all sensor nodes, with some nodes not considered to be off.

3. System Modeling

System modeling is a principal issue in studies of this nature and it needs to encompass various topics for a comprehensive understanding of what we have, what we present, and the preferences.

In fact, it is a basic concept that we need in order to evaluate past and present methods. Corresponding models should be presented and, accordingly, other related topics will be represented around them.

An industrial network can include three kinds of components: sensors, actuators, and central controllers, which are enumerated as corresponding components for traditional networks. The ordinary connection method of IIoT involves both pub and sub models. As an example of data sources, sensors can be defined and transmit data to a central controller; this component can store the data so that they are available to the actuators when they request it. In smart factories, where industrial network applications are subject to time constraints, access latency is of considerable importance, in accordance with caching relative data in the central controller from the consumers of actuators. Therefore, access latency is very important and requires special attention. On the one hand, the latency corresponds to data access and is important for numerous reasons, including the extensive distance between data and the central controller. On the other hand, the overhead surrounding the central controller can be attributed to the burden of highlighting, maintaining, and processing all network data through the central controller. Both traditional pub and sub models endure important and critical challenges concerning network lifetime, due to the vital energy consumption surrounding the central controller in addition to data path triangularization.

The elapsed time from the start of data distribution to the earliest node losing its energy is defined as the lifetime of the network. The purpose of this paper is to maximize the lifetime of the network. By considering all available and possible paths for the data if the proxies are certain, a path is chosen that leads to the maximization of the lifetime. In our proposed method, we also use the off and on properties of nodes and we consider off nodes that are not used in the path. To achieve the objective of the problem, for each piece of data, we identify the possible paths and select the paths that meet the maximum delay restriction. For every one of these paths, we calculate the energy discharge on the path nodes. Therefore, if that path is active, we determine the node in the path with the minimum remaining lifetime. Among all possible paths, we finally choose the path that leads to the maximum remaining lifetime. In particular, for every path, a node is considered to be the first to die in the network while that path is active; in this way, a path is selected wherein the nodes have the longest lifespan. In this article, the available data are prioritized and a group of data is considered urgent data. Urgent data are available to the consumer faster than other data, and the data access latency is analyzed in two separate groups of urgent and normal data. Since one of the goals of this plan is to reduce the data access latency, it is demonstrated that in the case of a semi-determined proxy, the amount of data access latency is reduced compared to a determined proxy; as a result, it improves network performance.

The models can be organized as follows. Their corresponding details are expressed below. In the model of the proposed system, internet devices of industrial objects are connected with each other, with the help of wireless communication links. We illustrate this in Figure 1. Some of the nodes in the network are producer nodes (sensors), some are consumer nodes (operators), and others are proxies.

In order to tackle the challenges mentioned earlier, we propose a system model, as depicted in Figure 1. In this model, certain components of the IIoT network act as proxy nodes, which are responsible for caching the data generated by the sensor nodes. This caching mechanism enables efficient data access. To ensure seamless data retrieval, each actuator is assigned to a suitable proxy node that holds the relevant cached data. By intelligently selecting proxy nodes and appropriately designating actuators to them, we guarantee that the data access latency remains below a predefined maximum threshold. This optimization not only improves performance but also minimizes the energy consumed during the data transmission between the sensor nodes and proxy nodes, as well as between the proxy nodes and actuators. A crucial element within our system model is the central controller, which assumes a managerial role by executing the EDMM. This scheme oversees the overall operation, coordination, and management of the network components, ensuring efficient data handling and resource allocation. Overall, our proposed system model, with its selection of proxy nodes, actuator assignments, and central controllers with the EDMM, aims to optimize data access latency and conserve energy in the industrial IoT environment.

All these nodes are connected to each other by means of communication links.

Suppose that $G = (V, E)$ is as a graph of an industrial IoT network, where V denotes a set of nodes of a graph G and every node $u \epsilon V$ has a limited amount of energy that can be defined as E_u. The network is able to characterize two kinds of nodes: resource-constrained sensors and data nodes in addition to potential proxy nodes that are placed in a set P. If P is the total number of proxy nodes, V is the number of nodes, and E_p is the limited amount of energy of proxy p, then $P \subset V$, $|P| \ll |V - P|$,and $E_p \gg E_u$, $\forall u \epsilon V$, $p \epsilon P$. A node $u \epsilon V$ can propagate data utilizing appropriate industrial wireless technologies to nodes that are in the neighborhood N_u. N_u includes nodes $v \epsilon V$ that satisfy $\gamma . \rho_u \geq \delta(u, v)$ so that ρ_u is the transportation limit area of node u, $\delta(u, v)$ is the Euclidean interval among u and v, and γ is a neighborhood adjustment parameter, where $0 < \gamma \leq 1$.

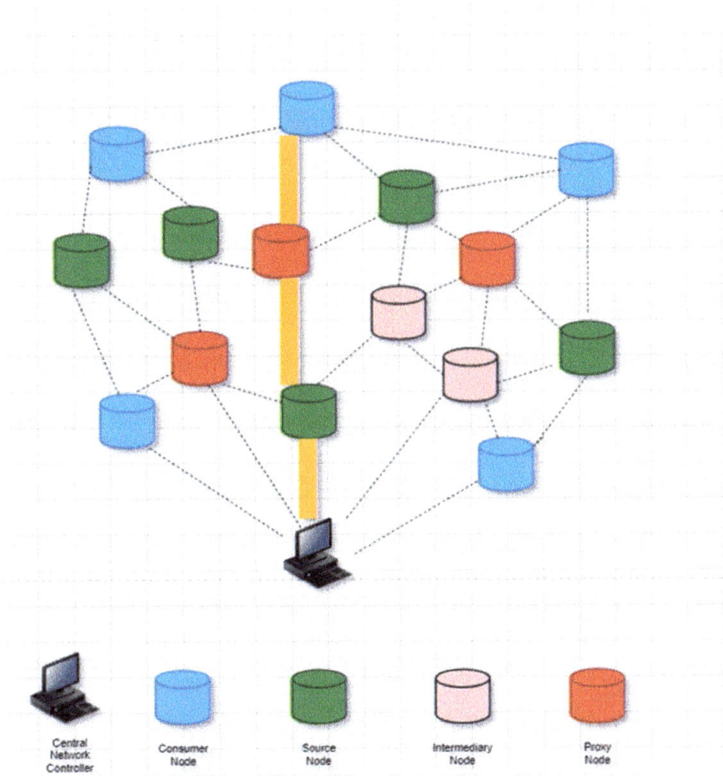

Figure 1. The proposed system model.

One essential aspect of industrial operations involves consumer access to data on demand (typically in a timely manner). A delivery system must ensure compliance with certain maximum data access latency constraints. l_{uv} is defined as a delay that includes one hop from u to v. The latency resulting from multiple hops, achieved from u to p, is defined by $L_{up} = l_{uv_1} + \ldots + l_{v_i p} + l_{pv_i} + \ldots + l_{v_1 u}$. It is shown in Figure 2.

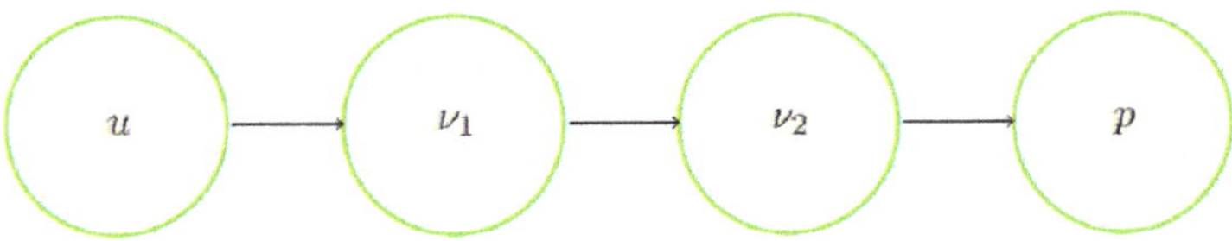

(**a**)

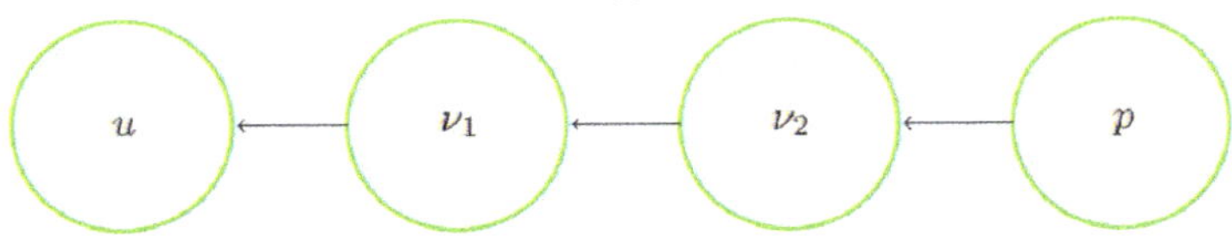

(**b**)

Figure 2. Data access delay. (**a**) Data request, (**b**) Data delivery.

The data access latency in Figure 2 is

$$L_{up} = l_{uv_1} + l_{v_1 v_2} + l_{v_2 p} + l_{pv_2} + +l_{v_2 v_1} + l_{v_1 u}.$$

Upon a requisition from c_i, data piece D_i is delivered from p through a (distinct) multi-hop path; the data access latency of c_i can be defined by

$$L_{c_i} = l_{c_i u} + \ldots + l_{vp} + l_{pv} + \ldots + l_{uc_i}$$

Urgent data with high-priority data parts should be sent quickly. Therefore, we consider $L_{\max}$ as the maximum tolerable delay for normal data and $L^u_{\max}$ as the maximum tolerable delay for urgent data, with $L^u_{\max} < L_{\max}$.

In some cases, data generation takes place in networks related to industrial processes. In general, data are divided into two groups: urgent data and normal data. Urgent data are data that are necessary to exist in the network. The data are introduced by D, where $D = \{D_1, D_2, \ldots, D_m\}$. Any data piece can be defined by $D_i = (s_i, c_i, u_i, r_i)$, where $s_i \epsilon V$ is the source of $D_i, c_i \epsilon V$ is the consumer of D_i, r_i represents the data production rate of D_i, where $i = 1, 2, \cdots, m$ and m is the number of data.

Given this constraint and the constraints that we will demonstrate in the following, the main aim for each data source, s_i, is the proxy recognition p, where the relevant data should be cached, for the purpose of maximizing the lifetime of the network. For the following topics, we are going to provide a suitable showcase for our main problem, i.e., the maximization issue. The purpose of modeling is to progress, and decisions should be made corresponding to our model representation Decision-making will be given in the next subsection.

Decision Problem

As mentioned previously, we should deal with the problem of what we can do. Clearly, our decision should be made and our constraints should be highlighted and explained. Regarding decision-making, many related issues are better clarified and understood. Meanwhile, there are numerous constraints that must be accepted, and considering such items, we want to choose the best options. However, the choice must be optimized and have at least one preference due to the other items. For the following sub-section, we will prove suitable decision constraints and their related decision-making procedures to achieve our goal. Further topics and complimentary topics will be presented.

Suppose that there is a set of deployed proxies p for a provided network $G = (V, E)$. Two situations are considered for each P: active (that is, communicating or caching) and idle. In the idle mode, P is ON but its internal storage is not in use and refuses to participate in sending or receiving data. Therefore, e^{on}_p denotes the energy costs of activating node P as a proxy node. Energy consumption costs can be defined by ε_{uv} for every $u, v \,\epsilon$ V.

The aim is to maximize network lifetime, which can be challenging in industrial IoT. Consequently, the time span from the initiation of data distribution to the moment when the first node in the network depletes its energy is defined as the lifetime of the network. To construct the objective function that maximizes the lifetime of the network, we present the decision variables, x^i_{uv}, which keep the essential information about the transport of the data pieces across the edges of the graph. In particular, $x^i_{uv} = 1$ when an edge (u, v) is activated for the data piece D_i. We denote $a_{uv} = \sum_{i=1}^{m} r_i x^i_{uv}$ as the sum of the data rate of (u, v). All a_{uv} is defined by $x = [a_{uv}]$. According to the above statements, the lifetime of node $u \epsilon V$ is given by

$$T'_u(X) = \frac{E_u}{e^{on}_u \sum_{v \epsilon N_u} (a_{uv} \varepsilon_{uv} + (\sum_{i=1}^{m} x^i_{uv}))} \tag{1}$$

The original objective function for this study, which can be enumerated as network lifetime, can be formulated as follows:

$$T'(X) = \min_{u \epsilon V} \{T'_u(X) | \sum_{v \epsilon N_u} x^i_{uv} > 0\} \tag{2}$$

4. Energy-Aware Data Management Method

The first problem is to find suitable objective issues that can be regarded in the decision-making process and formulation of our constraints. Among the different subjects, energy is one of the most important, applicable, and interesting topics to deal with. The energy is of interest in both internal and existing problems. There is an interest in minimizing internal energy consumption, in direct contrast to maximizing the available external energy. Having information about energy enables us to design a decision problem that can be useful in data management, utilizing energy amounts and energy-aware concepts. To address this, an energy-aware data management strategy, encompassing both theoretical and applied topics, will be presented. Methods for resolving these concerns are also discussed.

Here, we introduce the EDMM method that chooses proxy nodes from set P. This strategy simultaneously divides them into data pieces, to maximize network lifetime. In the same direction, we take into account the access delay and storage capacitance. We propose an algorithm to solve our problem of maximizing the lifetime in the network.

$$max : T^{'}(X) \tag{3}$$

$$s.t. L_{c_i}.x^i_{pc_i} \leq (1-u_i)L_{\max} + u_i L^u_{\max} \qquad \forall d_i \epsilon D, \quad p \epsilon P \tag{4}$$

$$\sum_{p \epsilon P} x^i_{pv} \geq 1 \qquad \forall i \tag{5}$$

$$\sum_{v \epsilon V} (x^i_{uv} - x^i_{vu}) = 0 \qquad \forall u \epsilon V \setminus s_i, c_i \tag{6}$$

$$\sum_{v \epsilon V} (x^i_{s_i v} - x^i_{v s_i}) = 1 \qquad \forall s_i \epsilon V \tag{7}$$

$$\sum_{v \epsilon V} (x^i_{c_i v} - x^i_{v c_i}) = -1 \qquad \forall c_i \epsilon V \tag{8}$$

$$\sum_{v \epsilon V} \sum_{i} \varepsilon_{uv} r_i x^i_{uv} \leq E_u \qquad \forall u \epsilon V \tag{9}$$

$$\sum_{v \epsilon V} x^i_{uv} \leq 1 \qquad \forall u \epsilon V, \forall i \tag{10}$$

$$x^i_{uv} \epsilon \{0,1\} \tag{11}$$

The constraints mentioned are briefly stated. Constraint (4) ensures that neither normal nor urgent data can exceed the latency thresholds. Constraint (5) ensures that one or more proxies are involved in the distribution of the data pieces. Data flow conservation is assured according to constraints (6)–(8) for all nodes. Regarding constraints (9), it is clear that the total energy consumption related to each node u will not exceed the primary level E_u. In the following, constraint (10) is able to make sure that any data piece is propagated from u through just one edge (u, v). Variables x^i_{uv} (11) are set to be integers that are understood based on the formulation of the problem.

For better understanding, the procedure algorithm will be presented. According to this algorithm, the strategy can be completely executed. The input and output can be clearly found, allowing us to base our intuition on the results Algorithm 1 demonstrates the procedure for finding our problem, which is obtained using the CPLEX tool.

Algorithm 1 EDMM.

network graph $G(V, E)$, set of data pieces D, energy of an active node e^{on}, limited energy node E_u, energy consumption costs ε_{uv}, $L_{\max}$, and $L^u_{\max}$ For all $d_i \epsilon D$
For all $p \epsilon P$
$D' \leftarrow$ Sort D from highest to the lowest r_i
$T'_u(X) \leftarrow$ Compute the lifetime by Equation (2)
$X \leftarrow$ Proxy for every data piece maximizing the objective function
X

5. Performance Evaluation

In this section, the performances of the presented model are tested and the corresponding results are evaluated. The processes were executed according to software procedures, and the related steps will be clearly defined. As it seems logical, firstly, we focus on setting the parameters and assigning their initial values. The analysis topics are then presented and the corresponding sensitivity analysis is explained. In addition, complementary algorithms, analyses, and comprehensive discussions will be covered. The advantages of our method will be clearly expressed in both algorithmic and software outputs. MATLAB is the software utilized in this part, and the corresponding results along with their detailed discussions can be found in this section.

Considering various network scales, we are going to demonstrate the efficiency of the proposed method, which we denote as EDMM, via an extensive assessment. Furthermore, we will compare the given strategy with the presented approach [5]. The analytical behaviors of the given model are based on the optimization issues by the CPLEX solver utilizing a MATLAB simulator.

5.1. Parameter Settings

The first step in evaluating the performance of our algorithm involves establishing initial values to enable the running of the algorithm. The initial parameters play an important role in the sensitivity analysis. It is better to choose them appropriately so that we can evaluate the performances and behaviors of the presented strategy. The choice of the initial parameter is expressed and further concepts are discussed.

In this part, a real application problem will be discussed. Consider the Inria Lille - Nord Europe, which hosts an Euratech testbed, consisting of a showroom and 224 nodes, as arranged below: A total of 2 horizontal forms are placed in a grid constitution of 5×19 nodes and 34 nodes are affixed to the wall, 0.60 m away from it. For our current purpose and to consider a plausible indoor industrial topology, we will consider 18 of the testbed nodes; this selection results in node distances ranging between 1.2 and 1.7 m. Assuming different existence power levels, we aim for a transmission power of 3 m, with $\gamma = 0.6$; this results in an average neighborhood consisting of five neighboring nodes, on average, where $v \epsilon N_u$ and where $d(u.v) \leq 2$ [5]. The proportion of data varies from 10% to 50% regarding the number of nodes. For any status, 50% of the data are categorized as urgent, and the remainder is designated as normal.

For the present examination, we utilize the WSN430 data that are provided in [5]. These data are mote-constructed data featuring a low-power MSP430-based structure equipped with standard sensors Further information can be found in [16]. It is worth mentioning that these data also engage with the IEEE 802.15.4 radio interface at 2.4 GHz. The antenna TX power has been set at −25 dBm, in alignment with the CC2420 antenna data [16]. We obtain our favorite scope, $\rho_u = 3$ m. The corresponding nodes have a maximum capacity of 830 mAh at 3.7 V and are battery-operated. Further facts and pieces of information that are based on simulation studies are revealed in Table 1.

Table 1. Experimental parameters.

Parameter	Value
Topology	
(2D grid)	2.4 m × 6.0 m
Number of nods	18
∣p∣	different numbers
∣s∣, and ∣c∣	different numbers
Hardware	
MCU	MSP430
Antenna	CC2420
Max. battery capacity	830 mAh, 3.7 V
E_u, and E_p	0–1, 3 Wh
Transmission power	−25 dBm
Energy of keeping the active node (e^{on})	6 mW
Time	
One-hope latency ($l^{(h)}$)	28 ms
($L_{\max}$) for normal	120
($L_{\max}$) for urgent	100
Data	
(D)	10–45% of ∣V∣
(u)	50% of (D)
Data piece generation rate r_i	1–8 D_i/s
Data piece size	9 bytes

5.2. *Analysis*

The second item that we investigate revolves around performing a comprehensive analysis of the efficacy of our proposed model We will present three figures to help elucidate the behaviors of our model.

To improve our evaluation, we compare EDMM with the approach presented in [5], where the authors introduced an offline centralized heuristic algorithm to assess their PDD. We compare EDMM and PDD, focusing on various evaluation criteria, such as network lifetime and access latency.

Network lifetime: We run the proposed approach (EDMM) in the network and consider c_i and r_i, as they increase from 1 to 8; the results are compared with an offline centralized heuristic algorithm (PDD). We illustrate the comparison in Figure 3. It is evident that an increase in the number of consumers results in a reduction in network lifetime. It is obvious that the EDMM is more efficient in comparison with PDD and the lifetime of the network obtained through method EDMM is longer than the PDD method.

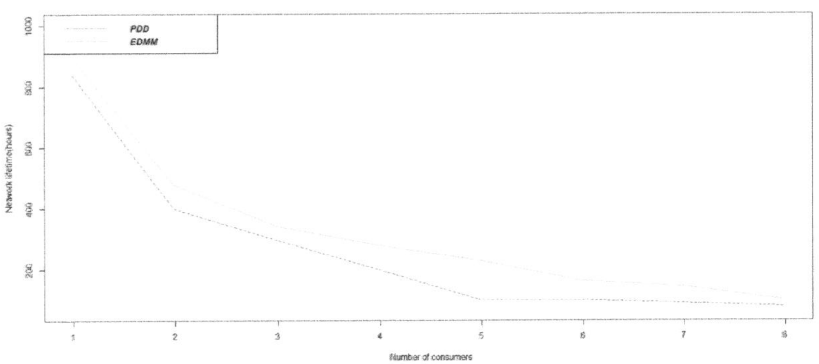

Figure 3. Network lifetime.

Data access delay: Data access delay has been studied and is determined by evaluating all consumers present in a network. The measure can be defined by the individual demands of the consumers to the related proxies that store their data, and it refers to asynchronous latency. For both urgent and normal data, EDMM ensures that the access latency is consistently below their corresponding maximum latency thresholds The EDMM ensures that, in this case, access latency remains below their corresponding maximum latency thresholds for both urgent and normal items. In this regard, and for a comparative illustration, the comparison between latency and urgent data can be found, respectively, in Figures 4 and 5. These figures show that the amount is divided fifty-fifty between these data. The urgent access latency and urgent data are compared in Figure 4. Accordingly, access latency is lower than 100 ms, and it is recognized for urgent data as the maximum latency threshold. It can be highlighted that our method, named EDMM, consistently remains below the maximum threshold for urgent data in all cases. Meanwhile, the alternative strategy, PDD, is unable to achieve the same. Finally, provide a measure combining both normal and urgent latency, termed as the total data access latency, where their rate of combination is equal. For a comprehensive and collaborative showcase of this statement, see Figure 5. The red line for both Figures 4 and 5 is the normal level for comparison with others.

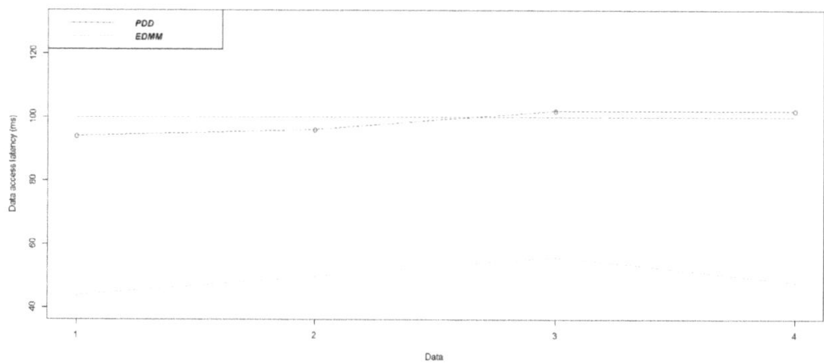

Figure 4. Urgent data access latency.

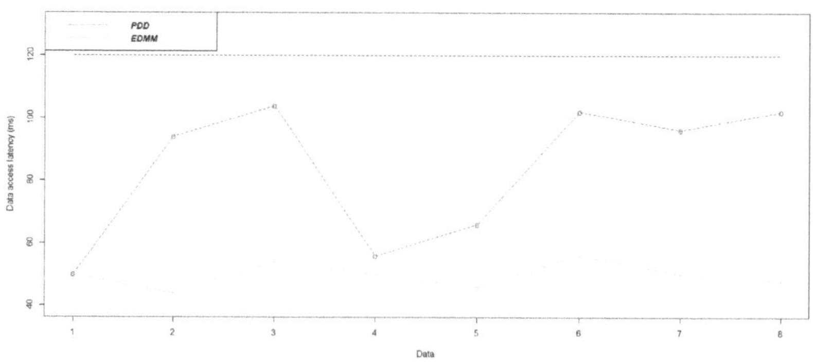

Figure 5. Overall data access latency.

6. Conclusions

In this paper, we present an energy-aware data management method (EDMM). In this method, a set of IoT nodes is chosen to store data in a distributed manner to maximize network lifetime. In addition, for the purpose of increasing battery lifetime, several nodes that do not participate in data transmission or storage are considered to be idle or off. The data available in the network are divided into two groups: urgent data and normal data. Accordingly, the access latency will be different for each group. Therefore, the maximum latency is considered for two groups, and the maximum latency for urgent data and normal data should not exceed the average latency of data access from these two thresholds. We illustrate that the proposed approach (1) ensures data access latency remains below a specified threshold, and (2) exhibits commendable network lifetime performance in comparison to an offline centralized heuristic algorithm. The maximum lifetime of the network in both directions, from the producer node to the proxy node and also from the proxy node to the consumer node, can be explored as future goals. It is also possible to consider the maximum lifetime of the network and to improve the issue; machine learning and deep learning mechanisms can also be used.

Author Contributions: The authors confirm their contribution to the paper as follows: study conception and design: Y.A.; data collection: Y.A.; analysis and interpretation of results: Y.A. and Z.M.; draft manuscript preparation: Y.A. and Z.M. All authors reviewed the results and approved the final version of the manuscript.

Funding: This research received no external funding.

Data Availability Statement: Not applicable.

Conflicts of Interest: The authors declare no conflict of interest.

References

1. Da Xu, L.; He, W.; Li, S. Internet of things in industries: A survey. *IEEE Trans. Ind. Inform.* **2014**, *10*, 2233–2243.
2. Raptis, T.P.; Passarella, A.; Conti, M. Distributed path reconfiguration and data forwarding in industrial IoT networks. In Proceedings of the International Conference on Wired/Wireless Internet Communication, Boston, MA, USA, 18–20 June 2018; Springer: Berlin/Heidelberg, Germany, 2018; pp. 29–41.
3. Shah-Mansouri, V.; Wong, V.W. Distributed maximum lifetime routing in wireless sensor networks based on regularization. In Proceedings of the IEEE GLOBECOM 2007-IEEE Global Telecommunications Conference, Washington, DC, USA, 26–30 November 2007; IEEE: New York, NY, USA, 2007; pp. 598–603.
4. Salman, T.; Jain, R. A survey of protocols and standards for internet of things. *arXiv* **2019**, arXiv:1903.11549.
5. Raptis, T.P.; Passarella, A.; Conti, M. Maximizing industrial IoT network lifetime under latency constraints through edge data distribution. In Proceedings of the 2018 IEEE Industrial Cyber-Physical Systems (ICPS), Saint Petersburg, Russia, 15–18 May 2018; IEEE: New York, NY, USA, 2018; pp. 708–713.
6. Raptis, T.P.; Passarella, A.; Conti, M. Performance analysis of latency-aware data management in industrial IoT networks. *Sensors* **2018**, *18*, 2611. [CrossRef] [PubMed]
7. Huo, Y.; Lin, X.; Di, B.; Zhang, H.; Hernando, F.J.L.; Tan, A.S.; Mumtaz, S.; Demir, O.T.; Chen-Hu, K. Technology Trends for Massive MIMO towards 6G. *Sensors* **2023**, *23*, 6062. [CrossRef] [PubMed]
8. Saqlain, M.; Piao, M.; Shim, Y.; Lee, J.Y. Framework of an IoT-based industrial data management for smart manufacturing. *J. Sens. Actuator Netw.* **2019**, *8*, 25. [CrossRef]
9. Kaur, K.; Garg, S.; Kaddoum, G.; Bou-Harb, E.; Choo, K.K.R. A big data-enabled consolidated framework for energy efficient software defined data centers in IoT setups. *IEEE Trans. Ind. Inform.* **2019**, *16*, 2687–2697. [CrossRef]
10. Ghaderi, A.; Movahedi, Z. An energy-efficient data management scheme for industrial IoT. *Int. J. Commun. Syst.* **2022**, e5167. [CrossRef]
11. Ghaderi, A.; Movahedi, Z. Joint Latency and Energy-aware Data Management Layer for Industrial IoT. In Proceedings of the 2022 8th International Conference on Web Research (ICWR), Tehran, Iran, 11–12 May 2022; IEEE: New York, NY, USA, 2022; pp. 70–75.
12. Fitzgerald, E.; Pióro, M.; Tomaszewski, A. Network lifetime maximization in wireless mesh networks for machine-to-machine communication. *Ad Hoc Netw.* **2019**, *95*, 101987. [CrossRef]
13. Wu, C.; Gunatilaka, D.; Saifullah, A.; Sha, M.; Tiwari, P.B.; Lu, C.; Chen, Y. Maximizing network lifetime of WirelessHART networks under graph routing. In Proceedings of the 2016 IEEE First International Conference on Internet-of-Things Design and Implementation (IoTDI), Berlin, Germany, 4–8 April 2016; IEEE: New York, NY, USA, 2016; pp. 176–186.
14. Raptis, T.P.; Passarella, A.; Conti, M. Distributed data access in industrial edge networks. *IEEE J. Sel. Areas Commun.* **2020**, *38*, 915–927. [CrossRef]

15. Dattatraya, K.N.; Rao, K.R. Hybrid based cluster head selection for maximizing network lifetime and energy efficiency in WSN. *J. King Saud-Univ.-Comput. Inf. Sci.* **2022**, *34*, 716–726. [CrossRef]
16. Instruments, T. CC2420 datasheet, 2.4 GHz IEEE 802.15.4 / ZigBee-ready RF Transceiver. Available online: https://www.datasheetarchive.com/datasheet?id=66183dbafa491d9516661fe05ea7917183615d&type=P&term=cc2420 (accessed on 19 August 2023).

computers

MDPI

Article

Rendezvous Based Adaptive Path Construction for Mobile Sink in WSNs Using Fuzzy Logic

Omar Banimelhem * and Fidaa Al-Quran

Network Engineering and Security Department, Jordan University of Science and Technology, P.O. Box 3030, Irbid 22110, Jordan; fmalquran17@cit.just.edu.jo
* Correspondence: omelhem@just.edu.jo; Tel.: +962-2720-1000 (ext. 22465)

Abstract: In this paper, an adaptive path construction approach for Mobile Sink (MS) in wireless sensor networks (WSNs) for data gathering has been proposed. The path is constructed based on selecting Rendezvous Points (RPs) in the sensing field where the MS stops in order to collect the data. Compared with the most existing RP-based schemes, which rely on fixed RPs to construct the path where these points will stay fixed during the whole network lifetime, we propose an adaptive path construction where the locations of the RPs are dynamically updated using a Fuzzy Inference System (FIS). The proposed FIS, which is named Fuzzy_RPs, has three inputs and one output. The inputs are: the remaining energy of the sensor nodes, the transmission distance between the RPs and the sensor nodes, and the number of surrounding neighbors of each node. The output of FIS is a weight value for each sensor node generated based on the previous three parameters and, thus, each RP is updated to its new location accordingly. Simulation results have shown that the proposed approach extends the network lifetime compared with another existing approach that uses fixed RPs. For example, in terms of using the first dead node as a metric for the network lifetime, when the number of deployed sensor nodes changes from 150 to 300, an improvement that ranges from 48.3% to 83.76% has been achieved compared with another related approach that uses fixed RPs.

Keywords: WSN; mobile sinks; rendezvous points; fuzzy inference system; IoT; network lifetime

Citation: Banimelhem, O.; Al-Quran, F. Rendezvous Based Adaptive Path Construction for Mobile Sink in WSNs Using Fuzzy Logic. *Computers* **2023**, *12*, 66. https://doi.org/10.3390/computers12030066

Academic Editor: Paolo Bellavista

Received: 21 December 2022
Revised: 22 January 2023
Accepted: 15 March 2023
Published: 20 March 2023

1. Introduction

With the continuous advances in information technology accompanied by the introduction of Internet of Things (IoT) applications, Wireless Sensor Networks (WSNs), which are the core of the IoT-based systems, are now a main player in our daily lives [1]. In WSNs, the data that is generated by the sensor nodes is usually collected by a central node called the sink or the base station (BS). Traditional WSNs with a fixed BS pose a primary limitation, which is called an energy hole problem, and which results as a consequence of depleting the energy of the nodes close to the sink rapidly because of the overwhelming traffic transmitted by other sensor nodes far from the sink. This could result in affecting the WSN efficiency, such as partitioning the network into unconnected areas. Indeed, partitioning the network is considered a crucial issue, especially in large-scale networks.

Recently, the widespread use of mobile robots has opened the door to leverage mobile sinks in WSNs for performing many tasks, such as employing them as data collectors in WSNs [2]. In fact, using mobile sinks (MS) can significantly increase the balance between the nodes and thus extend the network lifetime. Furthermore, it can increase the coverage of the network by reaching isolated uncovered areas. However, in any mobile sink-based approach, the balance between power consumption and the delay of data collection is still a real challenge and an active research area [3]. Different schemes of moving sinks have been proposed by the research community. These schemes can be divided into three categories: random, controlled, and predefined movement strategies [4]. In random movement methods, the MS follows a random path in its movement to collect the data. The main limitation of using random movement methods is the buffer overflow problem. Moreover, uncontrolled movements extend the overhead for each node until finding the location of the new position of the mobile sink and, thus, increase the ratio of dropped packets. In controlled schemes, the speed and direction of the next MS destination are determined according to the network situation. For example, the areas where the sensor nodes have urgent information will be given priority over the other locations. In a predefined movement, the MS moves according to a known fixed or dynamic path, which is generated dynamically as a function of network parameters such as the energy of the sensor nodes and their locations. To reduce the energy consumption in predefined movement methods, MS visits each location near sensor nodes and therefore yields more energy balance. However, visiting each sensor node will pose delay limitations, such as a long path length as well as increasing the delay of data delivery.

To address this issue, Rendezvous Points (RPs) schemes have been proposed [5,6]. In RP-based schemes, specific locations are chosen in order to reduce data gathering delays and balance energy consumption amongst nodes. The challenging task in such schemes is how to select the suitable RPs that satisfy both the energy consumption and the delay requirements. Indeed, RPs influence the path that an MS will follow to gather the data from the sensor nodes. Typically, the path of the MS is established by applying the traveling salesman problem (TSP) on the chosen RPs. Hence, the position and the number of RPs play a major role in constructing a path, which balances energy consumption and delay requirements. Figure 1 shows an example of a WSN where the RP-based model is used for data collection by the MS. However, selecting fixed RPs during the whole network operation results in an imbalance of energy consumption among the nodes that are associated with the same RP and thus reduce the network lifetime. Therefore, updating the locations of the RPs dynamically during the network operation will improve the performance of the network efficiently. In this paper, an adaptive path construction approach is proposed, where the locations of the RPs are dynamically updated using a Fuzzy Inference System (FIS). Three inputs are used to determine the updated RPs' locations. These inputs are: the remaining energy of the sensor nodes, the transmission distance between the RPs and the sensor nodes, and the number of surrounding neighbors of each node.

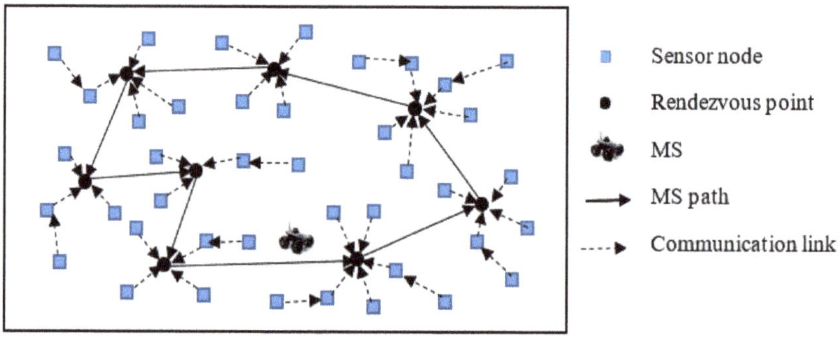

Figure 1. An example of a WSN where RP-based model is used for data collection by MS.

The remainder of the paper is structured as follows. The related work is presented in Section 2. The system model of the proposed approach is presented in Section 3. The proposed approach is discussed in Section 4. In Section 5, we present and discuss the simulation results, and the paper is concluded in Section 6.

2. Related Work

Several studies on random or controlled-based schemes for MSs have been conducted in WSNs ([7,8] are examples on random movement and [9,10] are examples of controlled-based movements). However, in this section, we address RP-based solutions that present a tradeoff between the random and controlled-based schemes in terms of the buffer flow problem in random movement schemes and the long delay problem in controlled-based schemes.

Park et al. [11] presented an approach in which the mobile sink moves along a fixed path and stops at a number of locations for collecting the data. The number of stop points over the path is selected using the Tabu search algorithm where the objective is minimizing the number of hop count from the sensor nodes to the mobile sink. Two algorithms called reduced k-means (RkM) and delay bound RkM (DBRkM) were proposed by Kaswan et al. [12] for generating a set of RPs that will be visited by the MS. The MS will then move over a fixed path that connects the selected RPs to gather the data from the sensor nodes. Banimelhem et al. [13] proposed an algorithm to generate a fixed path for the MS using principal component analysis (PCA), where the data can be gathered using either direct or multi-hop data transmission modes. A rendezvous-based routing protocol (RRP) was proposed by Sharma et al. [14] to address the need for energy efficiency and lower end-to-end latency. In the RRP, a rendezvous region is created in the middle of the network where the nodes, called backbone nodes, in this region form a tree, and where the other nodes communicate with the rendezvous region. Gupta et al. [6] proposed a routing method in WSN that depends on RPs and multiple MSs. At the beginning, the sensor nodes are distributed into a set of clusters using mean shift clustering (MSC). A cluster head (CH) for each cluster is then selected using the Bald Eagle Search (BES) algorithm. After that, the authors used the hybrid seagull optimization and salp swarm (SOSS) algorithm in order to find the RPs and the travelling route of each mobile sink in the network.

Vajdi et al. [15] proposed an approach that chooses a group of RPs outside the predetermined trajectory such that the defined path can accomplish the goals of minimizing sensor node energy consumption and decreasing network average data delivery time. Raj et al. [16] proposed an approach that builds a reliable and smart route for the mobile sink utilizing game theory and improves ACO-based MS route selection and the Data Gathering (GTAC-DG) algorithm. A set of rendezvous points (RPs) is selected to construct the path for the MS using an ACO-based algorithm. The GTAC-DG algorithm is used to create a path for the MS. Boyineni et al. [17] proposed an approach called the ant colony optimization (ACO)-based mechanism (ACO-RMS) for selecting the RPs and scheduling the mobile sink in the event-driven WSNs. The load of each sensor in the ACO-RMS approach is initially identified using a spanning tree. Different factors, such as distance, remaining energy, and total packets generated by the sensor nodes at a particular time, are used for the RPs' selection and path use. Donta et al. [18] proposed an extended ant colony optimization (ACO)-based MS path construction for event-driven WSNs, where the ACO algorithm selects the best set of RPs and the path that the MS will travel between these RPs.

Ghaleb et al. [19] proposed an approach where RPs are used to collect the data using data compression techniques from nearby sources and then send the data to the mobile sink when it travels over the path connecting these RPs. Furthermore, the authors proposed an algorithm called a minimal constrained rendezvous point (MCRP), which ensures that the collected data are relayed to the RPs taking into consideration three constraints: RPs' locations, bounded relay hop, and number of nearby sources.

3. System Model

In this paper, we propose an adaptive data collection scheme for homogeneous WSNs, where a set of sensor nodes S are deployed randomly in the sensing area. A single MS will move in the network area to collect the sensed data. We assume that the MS moves at a constant speed and it stops at RPs to collect the data. The stopping duration at each RP is assumed to be fixed and enough to collect the data from all corresponding sensors. For energy consumption, the energy model proposed in [20] is assumed. Using this model, the energy consumed (E_T) to transmit a k_b-bit packet between two sensor nodes s_1 at location (x_1, y_1) and sensor node s_2 at location at (x_2, y_2) is calculated as:

$$E_T = \left\{ \begin{array}{ll} k_b E_{elec} + k_b \varepsilon_{fs} d^2, & d < d_0 \\ k_b E_{elec} + k_b \varepsilon_{amp} d^4, & d \geq d_0 \end{array} \right\} \tag{1}$$

and the energy consumed by sensor node s_2 to receive the k_b-bit packet (E_R) is calculated as:

$$E_R = k_b\, E_{elec} \tag{2}$$

where E_{elec} is the electronics energy; ε_{fs} and ε_{amp} denote the amplifier energy that depends on the required receiver sensitivity and the receiver noise figure, respectively; and d is the distance between sensor nodes s_1 and s_2 and is given as:

$$d = \sqrt{(x_1 - x_2)^2 - (y_1 - y_2)^2} \tag{3}$$

The notations that are used in this paper are shown in Table 1.

Table 1. Notations used.

Term	Definition
S	Set of sensor nodes
s_i	Sensor node i
S_j	Set of sensor nodes associated with RP j
P	Set of potential RPs
C	Set of RPs that are used to build the MS path
$S_{RP}(i)$	Number of sensor nodes associated with RP i
$H_{RP}(i)$	Average hop distance between RP i and the sensor nodes associated with it
(X_{RPi}, Y_{RPi})	Location of RP_i
E_i	Remaining energy of sensor node i
$d_{i \to RPi}$	The distance between sensor node i and its corresponding RP
NBi	The 1-hop neighbors of sensor node i
PL	Path length
v	MS speed

4. Proposed Approach

In this section, we discuss the proposed approach for building a dynamic path for the mobile sink to collect the data from the sensor nodes in the network. The path is built based on selecting a set of RPs that will be used as stopping points where the MS will gather the data from the sensor nodes. First, we introduce the algorithm that determines the initial locations of the RPs, and we then discuss the algorithm that is used to update the RPs' locations in each round of data collection.

4.1. Initial Locations of the RPs

The initial locations of the RPs are obtained as given in Algorithm 1. First, as in [12], a set of P RPs is obtained by clustering the sensor nodes into $|P|$ clusters using k-means algorithm [21]. These $|P|$ points represent the set of candidate RPs for building the MS path. Each RP i, ($1 \le i \le |P|$) is then assigned a priority value $R(i)$ using Equation (4):

$$R(i) = \frac{S_{RP}(i)}{H_{RP}(i)} \tag{4}$$

where $S_{RP}(i)$ is the number of sensor nodes associated with RP i and $H_{RP}(i)$ is the average hop distance between RP i and the sensor nodes associated with it. Equation (4) gives high priority for the RP, which has more sensor nodes attached to it, and the average distance of these nodes to that RP is small compared to the other potential RPs. Assume PL is the length, in meters, of the path that connects all required RPs to collect the data. Assume the time that the MS needs to receive the data from the sensor nodes when it stops at each RP is T_{data}, then the total time for collecting the data from all sensor nodes T_{total} in each round is given as:

$$T_{total} = c \times T_{data} + \frac{PL}{v} \tag{5}$$

where c is the number of RPs obtained using Algorithm 1 ($c \leq |P|$) and v is the MS speed. Some WSN applications require that the sensed data should be collected within a specific delay limit. Therefore, depending on the WSN application (for example, a real-time application), the path length should not exceed a threshold value ($PL_{threshold}$). Based on that, in Algorithm 1, after determining the candidate set P as RPs using k-means algorithm, only c RPs of this set will be selected to build the MS path.

Algorithm 1: Finding the initial locations of the RPs

INPUT: *S*, $PL_{threshold}$
OUTPUT: *C, path for MS*
Begin *RPs INITIAL LOCATIONS*
P = k-means(S); // Cluster the S sensor nodes into P clusters using k-means algorithm [21]
for *i = 1 to* **P** *do*
    Calculate the priority value of RP i using Equation (4)
End for
Sort the set P of RPs based on their priority values in descending order
C = { }; / Set C contains the RPs that will be used to construct the MS path*/*
RPx = remove RP from P
$C = C \cup \{RPx\}$
c = 1
While True do
RPy = remove RP from P
    $C = C \cup \{RPy\}$
c = c + 1
    PL = TSP(C);/ Call traveling sales person algorithm to obtain the path between the RPs in C */*
    If $PL \geq PL_{threshold}$ ***then*** *break*
End While
If *(size(P) > 0)* **then** */* if some RPs in P are not used to construct the path */*
Redistribute the nodes attached to the RPs in P to the RPs in C
until the path length is equal or larger than the specified threshold.
End if
End

Once the initial RPs are selected and the initial path for data collection is constructed, the MS determines for each sensor a corresponding RP where each sensor node will be assigned to the closest RP. The MS then broadcasts a rendezvous information packet (RIP) to the whole network to inform each node about its RP. Therefore, when each sensor node receives the RIP packet, it obtains its destination RP. Once all sensor nodes know the destination RPs, the MS starts to accomplish data gathering by going through all RPs. When the MS becomes close to each destination RP, it sends a polling message with the ID of the corresponding RP. In this case, the corresponding sensor nodes prepare and transmit their data to the MS when it reaches the corresponding RP. If a node is within the MS communication range, it sends its data directly. Otherwise, it forwards the data to the nearest node of the corresponding RP. This procedure is repeated until all RPs are visited.

If the RPs are kept fixed during the entire operation of the network, it could pose energy holes around the fixed RPs location and, thus, the MS will no longer be able to receive data from sensor nodes with two or more hops around the RPs. Therefore, in this paper, a dynamic approach for updating the RPs locations every predefined number of rounds is proposed. For this purpose, we assume the MS keeps track of the following parameters during each round:

1. The current locations of the RPs.

2. The ID of the corresponding RP for each sensor.
3. The distance between each sensor and the corresponding RP.
4. The energy level of each sensor.
5. The number of surrounding sensor nodes for each sensor node.

Based on this information, the locations of the RPs are appropriately updated to extend the network lifetime and reduce the overall energy consumption as well. To perform the dynamic data collection for each new updated RPs' selection, the MS updates the ID of the corresponding RP of each sensor and broadcasts new RIP packets to inform the sensors with the new ID of the destination RP.

4.2. Updating the RPs Locations

In the proposed approach, the locations of the RPs are updated dynamically using a Fuzzy Inference System (FIS) [22]. An example of using fuzzy logic to control the MS movement toward cluster heads for improving the LEACH protocol [23] was proposed in [24]. In this paper, the proposed algorithm is called Fuzzy_RPs. In each round, the location of each RP is updated based on three factors:

1. The energy of each sensor around the RP. The energy level of each sensor influences how much the RP should be moved close to the sensor node. To balance the energy consumption and extend the network lifetime, the sensor with a low energy level has more impact to change the RP location and bring the RP close to it.
2. The distance between the sensor node and the corresponding RP. As the distance of transmission influences the amount of energy consumption, the location of RP should be updated to balance the distance between all sensor nodes and their corresponding RP. This factor has a significant impact to mitigate the overall energy consumption during the network lifetime.
3. The number of sensor nodes around each node. The sensor node within the dense region has more impact compared to the sensor node within the sparse region to influence the change of the RP location. The idea behind this factor is to attract the MS towards the dense region in order to be close to as many nodes as possible and therefore reduce the energy that will be used for transmission by the sensor nodes. This factor breaks the tie when two or more nodes have the same distance to their corresponding RP. For example, when two sensors have the same energy level and the same distance from the current corresponding RP, the sensor node with a high number of surrounding sensor nodes will attract the MS to update the location of the current corresponding RP towards it more closely compared with the sensor node in the sparse region.

The proposed FIS has three Inputs, which represent the three factors discussed above. For each sensor node *i*, the inputs of the FIS will be:

1. E_i: the remaining energy of sensor node *i*.
2. $d_{i \rightarrow RPi}$: the distance between sensor node *i* and its corresponding RP.
3. *NBi*: the 1-hop neighbors of sensor node *i*.

The output of the FIS determines a weight value wj between 0 and 1. The weight value increases if the distance is high or the energy of the node is low. The output weight value influences the change of the RP location. The sensor node with a higher weight value has more impact to change the location of the RP toward its location. Figure 2 shows the block diagram of the proposed Fuzzy_RPs. It consists of the basic four stages of FIS: fuzzification, evaluation of the fuzzy rules, aggregation, and deffuzzyfication of the output fuzzy sets where the Mamdani model is used for this purpose.

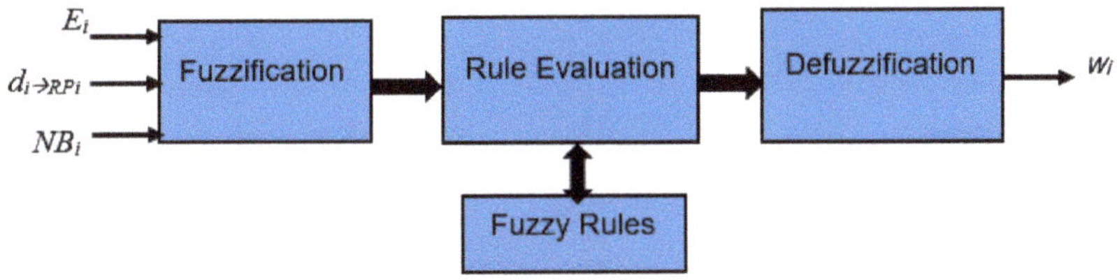

Figure 2. Block diagram of Fuzzy_RPs FIS.

The membership functions of the inputs and output are shown in Figure 3. The fuzzy rules are presented in Table 2, where AND operator is used to combine the three inputs. The shape of the membership functions of the fuzzy sets for the inputs and the output as well as the fuzzy rules are considered after running a set of simulations. In each run, the range of each fuzzy set and the fuzzy rules were evaluated, and the system was tuned until we got the membership functions shown in Figure 3 and the fuzzy rules given in Table 2.

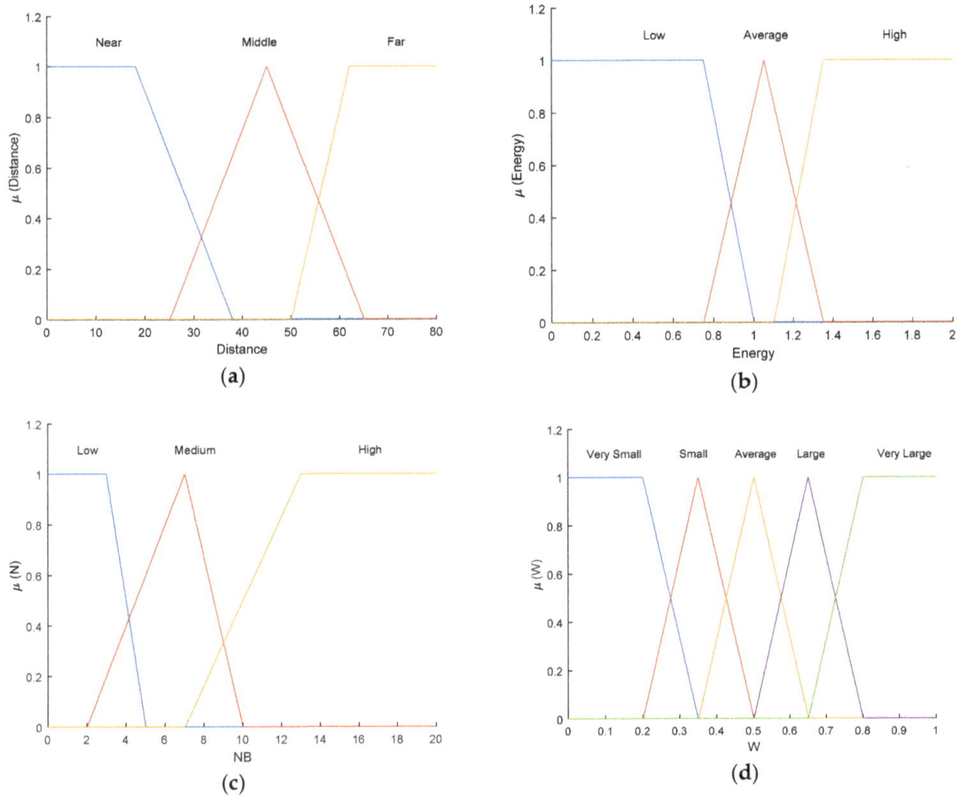

Figure 3. Membership functions of the inputs and output. (**a**) Membership functions of input Distance. (**b**) Membership functions of input Energy. (**c**) Membership functions of input (*NB*). (**d**) Membership functions gof output w.

Table 2. FIS Rules.

Rule Number	Inputs			Output w
	Distance	Energy	Number of Neighbors	
1	Near	Low	Low	Average
2	Near	Low	Average	Large
3	Near	Low	High	Very Large
4	Middle	Low	Low	Average
5	Middle	Low	Average	Large
6	Middle	Low	High	Very Large
7	Far	Low	Low	Large
8	Far	Low	Average	Very Large
9	Far	Low	High	Very Large
10	Near	Average	Low	Small
11	Near	Average	Average	Average
12	Near	Average	High	Large
13	Middle	Average	Low	Small
14	Middle	Average	Average	Average
15	Middle	Average	High	Very Large
16	Far	Average	Low	Average
17	Far	Average	Average	Large
18	Far	Average	High	Large
19	Near	High	Low	Very Small
20	Near	High	Average	Small
21	Near	High	High	Average
22	Middle	High	Low	Very Small
23	Middle	High	Average	Average
24	Middle	High	High	Small
25	Far	High	Low	Average
26	Far	High	Average	Large
27	Far	High	High	Large

Assume the current position of RP i is (X_{RPi}, Y_{RPi}), and the current position of sensor node j that is associated with RP i is (X_j, Y_j). Sensor node j will then attract RP i to its location and calculate its new position ($X_{RPi\to j}$, $Y_{RPi\to j}$) after calculating its weight value w_j using the FIS as follows:

$$X_{RPi\to j} = w_j \times (X_j - X_{RPi}) + X_j \tag{6}$$

$$Y_{RPi\to j} = w_j \times (Y_j - Y_{RPi}) + Y_j \tag{7}$$

The actual new updated position of RP i is based on all sensor nodes that are associated with RP i, and it is calculated as:

$$X_{Rpi}^{New} = \frac{\sum_{j=1}^{|S_j|} X_{RPi\to j}}{|S_j|} \tag{8}$$

$$Y_{Rpi}^{New} = \frac{\sum_{j=1}^{|S_j|} X_{RPi\to j}}{|S_j|} \tag{9}$$

where $|Sj|$ represents the total number of sensor nodes associated with RP i. After the locations of the RPs are updated, the MS constructs the adaptive path that passes through the new RPs set using Traveling Sales Man Problem Algorithm. Based on the new set of RPs, the mobile sink updates the corresponding RP of each sensor based on the nearest distance between each sensor with the new corresponding RP. When the mobile sink is about to reach an RP, it broadcasts a new polling message to the corresponding sensors with the ID of the new respective RP. Therefore, the sensors based on the polling message prepare the data and transmit it to the MS when it stops at the RP. Once the MS finishes the collection of the data from all corresponding sensor nodes, it moves to the next RP and repeats the process until it passes the whole set of RPs. Algorithm 2 presents the steps for updating the RPs' locations using FIS.

Algorithm 2: Fuzzy-based Adaptive Path Selection

INPUT: *S, C, U*
/* *U is the period value to update the path* */
/* *C is the set of RPs that are used to build the MS path* */
/* *S is the set of sensor nodes* */
OUTPUT: *updated path for MS*
Begin *FUZZY_RPs*
Round = 1; /* *current round of collection data* */
while *there is still active nodes* **AND** *(mod(Round, U) = 1) OR Round = 1)* **do**
for $j = 1$ *to sizeof(S)* **do**
Determine the Nearest Corresponding RP of sensor s_j
Discover number of the One-Hop nodes of sensor s_j
End for
for $i = 1$ *to sizeof(C)* **do**
$RP_i = Ci$
for *each sensor* s_j *associated with* RP_i **do**
w_j *=FIS [energy(*s_j*), dist(*s_j *to RPi), Num of one Hop nodes (*s_j*)*
Calculate $XRP_{i \to j}$ *using Equation (6)*
Calculate $YRP_{i \to j}$ *using Equation (7)*
End for
New X of RPi = Calculate New X_{RPi} *using Equation. (8)*
New Y of RPi Calculate New Y_{RPi} *using Equation (9)*
End for
Round = Round + 1
End while
Use TSP algorithm [25] to construct the path that passes through the RPs in C
End

5. Performance Evaluation

In this section, the performance of the proposed approach is evaluated and compared with the DBRkM approach [12]. The parameters that were used in the simulation experiments are shown in Table 3. Each experiment with the same configuration was repeated for 10 runs where in each run the same number of sensor nodes is deployed randomly in the network. The average value is then considered for the experiment. The performance metrics that were used in the simulation are: the number of active sensor nodes, total energy consumption, the standard deviation of the remaining energy, and the path length during the network lifetime.

Figures 4 and 5 show the area of the sensing field and the network conditions using the DBRkM approach and the proposed FUZZY_RPs approach, respectively. As shown in Figure 4, the sensor nodes that are associated with RPs 3 and 4 suffer from high hop count and large distance. On the contrary, Figure 5 shows that the locations of RP 3 and 4 are updated at round 175 to balance the distance between the sensor nodes towards the RPs and to reduce the average hop count. These two figures are presented to show that the RPs are built and changed dynamically in the FUZZY_RPs approach (Figure 5) while the RPs' locations are fixed in the DBRkM approach.

Table 3. Parameters used in the simulation.

Parameter	Value
Target Area	220 × 220 m^2
Number of sensor nodes	150–300
Initial Energy of sensor nodes	2 Joule
Communication Range (*Rc*)	40 m
Packet Size (K_b)	4000 bits
Speed of mobile sink (v)	2 m/s
E_{elect}	50 nJ/bit
Mp	0.0013 pJ/bit/m^4

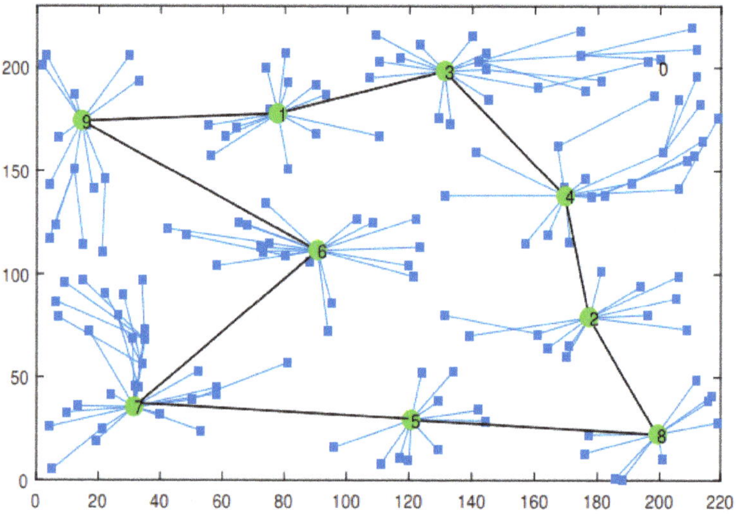

Figure 4. DBRkM Run-Time Simulation (average hop count = 1.24, tour length = 1034 m fixed at all rounds.

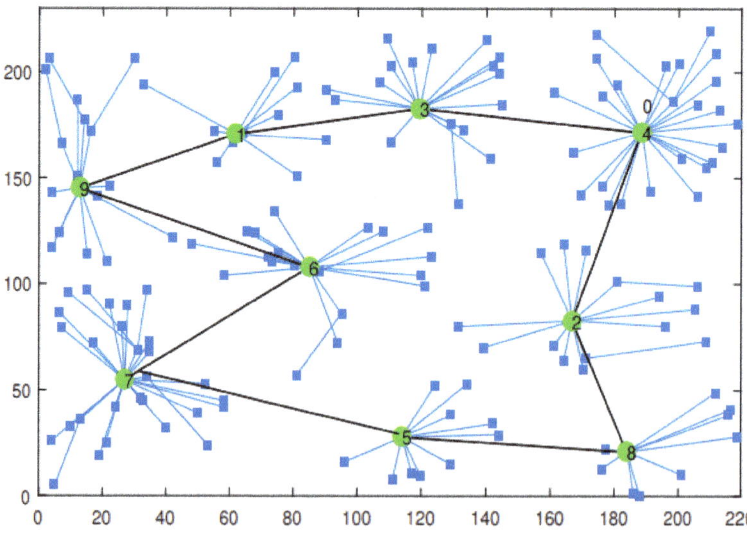

Figure 5. Fuzzy_RPs Run-Time Simulation at round 175 (average hop count = 1.12, tour length = 1016 m).

5.1. Network Lifetime

In this section, we compare the performance of the proposed approach with the DBRkM approach in terms of network lifetime. Three metrics are used for this purpose: the number of alive nodes, total remaining energy, and standard deviation of the remaining energy. Figure 6 shows the number of active nodes per round in both algorithms. The figure shows a significant improvement of the proposed Fuzzy_RPs approach compared to the DBRkM approach. This improvement is achieved by updating the locations of the RPs based on the energy of the nodes and therefore avoiding the energy holes around the RP.

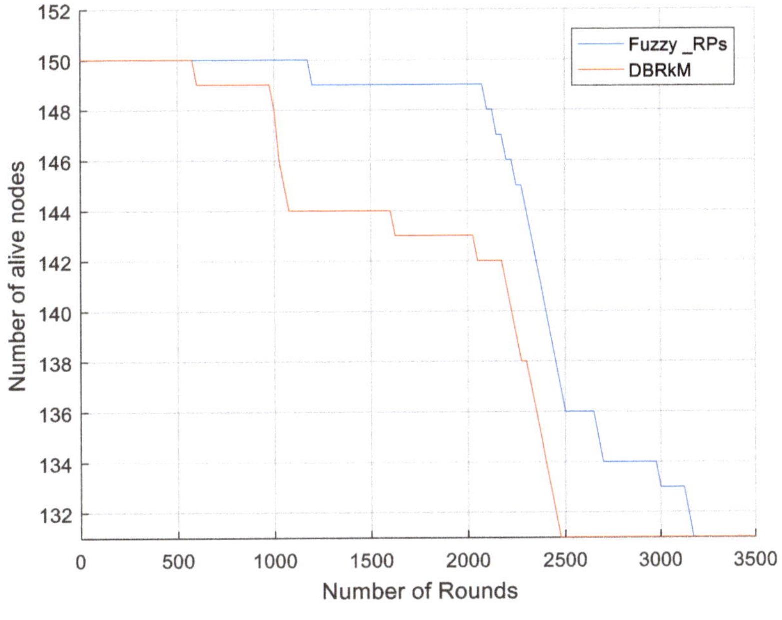

Figure 6. Number of alive nodes vs. rounds.

Figure 7 shows the overall amount of energy consumed throughout the first 3000 rounds. As can be seen in the figure, the proposed approach consumes less energy than the DBRkM approach. This improvement is achieved by keeping the RPs' points updated with the fewest hops and the shortest transmission distance possible. Figure 8 shows the standard deviation of the total remaining energy for live nodes per round for the proposed approach and the DBRkM approach. As shown in Figure 8 the standard derivation in the case of DBRkM approach increases compared to the Fuzzy_RPs approach. This increase comes from the fact that the RPs in DBRkM approach are fixed and therefore the nodes that are far away from these RPs will drain their energy fast compared to the nodes close to the RPs. On the contrary, the Fuzzy_RPs approach always achieves a balance in energy consumption by changing the RPs' locations in each round.

Figure 9 shows the network lifetime when the number of sensor nodes is changed from 150 through to 300. The network lifetime in this figure represents the number of rounds until the first node in the network dies. As shown in the figure, the proposed Fuzzy_RPs approach always outperforms the DBRkM approach. Table 4 shows the percentage of improvement that has been achieved by using the Fuzzy_RPs approach compared to DBEkM approach.

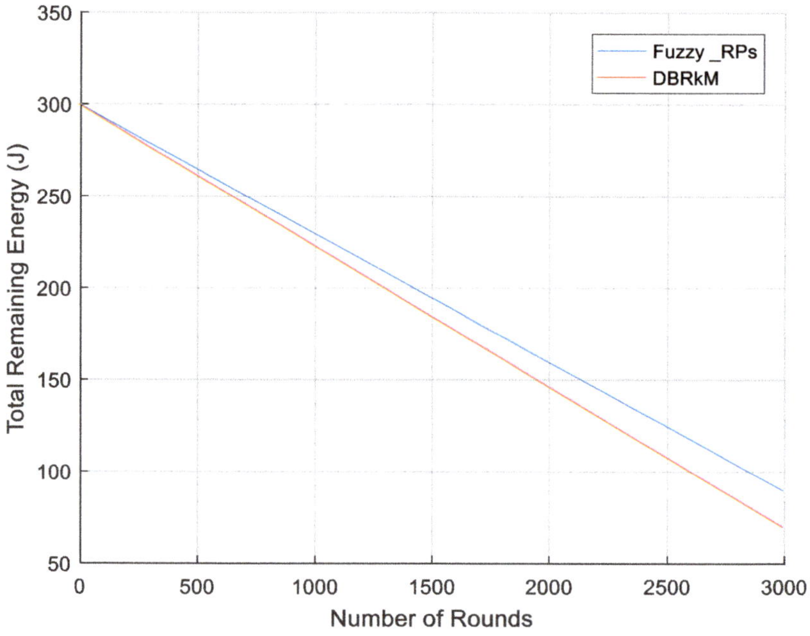

Figure 7. Total remaining energy vs. rounds.

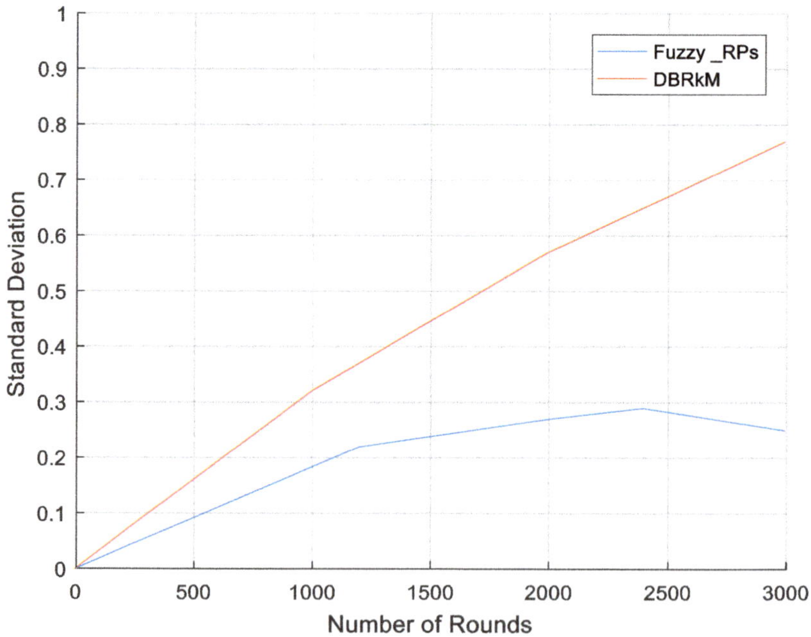

Figure 8. Standard deviation of the remaining energy vs. rounds.

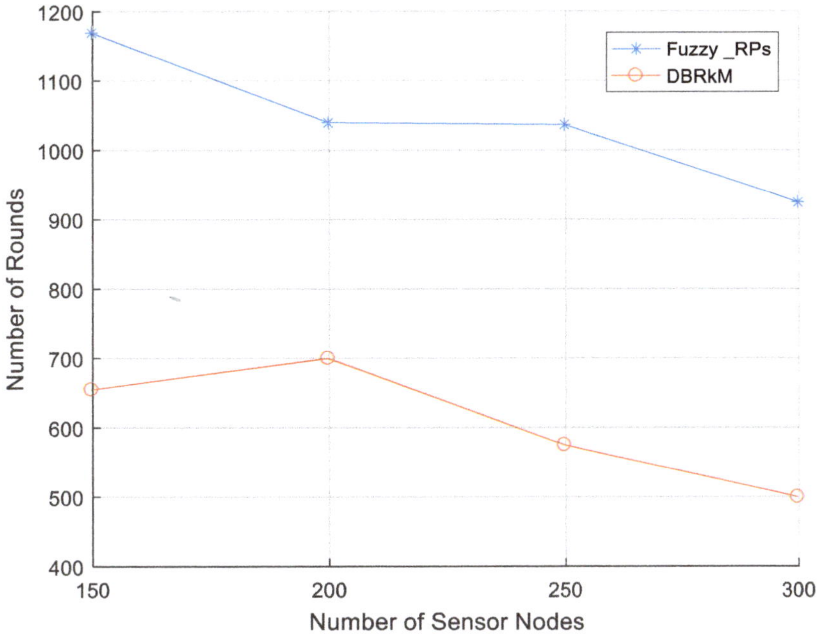

Figure 9. Network lifetime vs. number of sensor nodes.

Table 4. Network lifetime improvement with different number of sensor nodes.

Number of Sensor Nodes	Number of Rounds until First Node Dies		Improvement (%)
	Fuzzzy_RPs Approach	DBRkM Approach	
150	1172	655	78.93
200	1044	704	48.30
250	1037	575	80.35
300	928	505	83.76

5.2. Path Length during the Network Life Time

We assume that the MS energy is much more than the sensor node energy. The cost of energy dissipated as a result of the MS movement is represented by the length of the path that the MS will use to collect the data. Figure 10 shows the path length in meters in the DBRkM and FUZZY_RPs approaches. In this experiment, the number of sensor nodes was 150. As shown in the figure, the path length in the DBRkM approach is constant while the path construction is dynamic during the network lifetime in the FUZZY_RPs approach. As shown in the figure, the path length exceeds the DBRkM approach only in the last stage of the network, almost when the remaining energy in the network is low and most of the nodes are dead. As shown in Table 4, for the case when the number of sensor nodes is 150, the first dead node in the FUZZY_RPs approach occurs in round 1172 and for the DBRkM approach in round 655. However, during the lifetime of the network, the path length in the FUZZY_RPs approach is less than the path length in the DBRkM approach, especially in the first 2700 rounds.

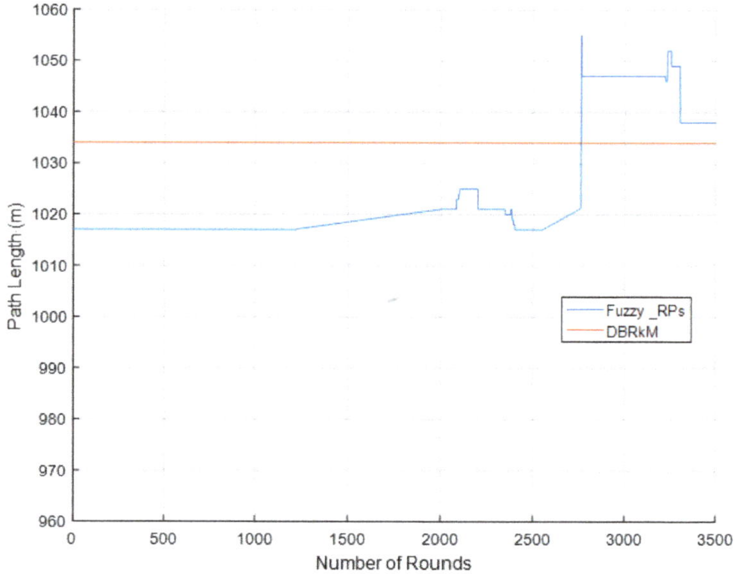

Figure 10. Path Length.

6. Conclusions

In this paper, we proposed an adaptive path construction for the mobile sink using FIS called FUZZY_RPs. After determining the required set of RPs using the k-means algorithm in the first round, the locations of these RPs are then adapted dynamically in the subsequent rounds using the FIS, which calculates the new location of each RP based on a weight value generated by each sensor node in the network. This weight value generated by a certain node determines how long the current location of the corresponding RP should be moved towards that node. The fuzzy-based calculation of the weight value for each sensor node depends on three factors: the remaining energy of the sensor node, the distance between the sensor node and its corresponding RP, and the 1-hop neighbors of the sensor node. Simulation results showed that the proposed approach outperforms the DBLkM approach, which uses fixed RPs over the whole network lifetime. For example, in terms of the network lifetime, the proposed FUZZY_RPs approach achieved an 83.76% improvement when the number of sensor nodes is 300. Future work can look into the adaptive path design for the MS by updating the locations of the RPs in environments with obstacles.

Author Contributions: Implementation, software, validation, writing—original draft preparation, data collection, and analysis, F.A.-Q.; supervision, writing—review and editing, project administration, O.B. All authors have read and agreed to the published version of the manuscript.

Funding: This research was funded by Jordan University of Science and Technology grant number 20200241.

Informed Consent Statement: Not applicable.

Data Availability Statement: The data used to support the findings of this study are available from the corresponding author.

Conflicts of Interest: The authors declare no conflict of interest.

References

1. Abdur, R.M.; Arafatur, R.M.; Rahman, M.M.; Asyhari, A.; Bhuiyane, M.; Ramasamy, D. Evolution of IoT-enabled connectivity and applications in automotive industry: A review. *Veh. Commun.* **2021**, *27*, 100285. [CrossRef]
2. Agarwal, V.; Tapaswi, S.; Chanak, P. A Survey on Path Planning Techniques for Mobile Sink in IoT-Enabled Wireless Sensor Networks. *Wirel. Pers. Commun.* **2021**, *119*, 211–238. [CrossRef]
3. Akbar, M.; Javaid, N.; Abdul, W.; Ghouzali, S.; Khan, A.; Niaz, I.A.; Ilahi, M. Balanced Transmissions Based Trajectories of Mobile Sink in Homogeneous Wireless Sensor Networks. *J. Sens.* **2017**, *2017*, 4281597. Available online: https://www.hindawi.com/journals/js/2017/4281597/ (accessed on 1 September 2022). [CrossRef]
4. Temene, N.; Sergiou, C.; Georgiou, C.; Vassiliou, V. A survey on mobility in wireless sensor networks. *Ad Hoc Netw.* **2022**, *125*, 102726. [CrossRef]
5. Suh, B.; Berber, S. Rendezvous points and routing path-selection strategies for wireless sensor networks with mobile sink. *Electron. Lett.* **2016**, *52*, 167–169. [CrossRef]
6. Gupta, P.; Tripathi, S.; Singh, S. Energy efficient rendezvous points based routing technique using multiple mobile sink in heterogeneous wireless sensor networks. *Wirel. Netw.* **2021**, *27*, 3733–3746. [CrossRef]
7. Irish, A.E.; Terence, S.; Immaculate, J. Efficient data collection using dynamic mobile sink in wireless sensor network. In *Wireless Communication Networks and Internet of Things, Lecture Notes in Electrical Engineering*; Springer: Singapore, 2019; pp. 141–149. [CrossRef]
8. Wang, J.; Gao, Y.; Yin, X.; Li, F.; Kim, H. An enhanced PEGASIS algorithm with mobile sink support for wireless sensor networks. *Wirel. Commun. Mob. Comput.* **2018**, *2018*, 9472075. [CrossRef]
9. Liang, W.; Luo, J.; Xu, X. Prolonging network lifetime via a controlled mobile sink in wireless sensor networks. In Proceedings of the IEEE Global Telecommunications Conference (GLOBECOM), Miami, FL, USA, 6–10 December 2010; pp. 1–6. [CrossRef]
10. Farzinvash, L.; Najjar-Ghabel, S.; Javadzadeh, T. A distributed and energy-efficient approach for collecting emergency data in wireless sensor networks with mobile sinks. *AEU Int. J. Electron. Commun.* **2019**, *108*, 79–86. [CrossRef]
11. Park, J.; Moon, K.; Yoo, S.; Lee, S. Optimal stop points for data gathering in sensor networks with mobile sinks. *Wirel. Sens. Netw.* **2012**, *4*, 8–17. [CrossRef]
12. Kaswan, A.; Nitesh, K.; Jana, P.K. Energy efficient path selection for mobile sink and data gathering in wireless sensor networks. *AEU-Int. J. Electron. Commun.* **2017**, *73*, 110–118. [CrossRef]
13. Banimelhem, O.; Taqieddin, E.; Shatnawi, I. An Efficient Path Generation Algorithm Using Principle Component Analysis for Mobile Sinks in Wireless Sensor Networks. *J. Sens. Actuator Netw.* **2021**, *10*, 69. [CrossRef]
14. Sharma, S.; Puthal, D.; Jena, S.K.; Zomaya, A.Y.; Ranjan, R. Rendezvous based routing protocol for wireless sensor networks with mobile sink. *J. Supercomput.* **2017**, *73*, 1168–1188. [CrossRef]
15. Vajdi, A.; Zhang, G.; Zhou, J.; Wei, T.; Wang, Y.; Wang, T. A New Path-Constrained Rendezvous Planning Approach for Large-Scale Event-Driven Wireless Sensor Networks. *Sensors* **2018**, *18*, 1434. [CrossRef] [PubMed]
16. Raj, P.V.P.; Khedr, A.M.; AL Aghbari, A. Data gathering via mobile sink in WSNs using game theory and enhanced ant colony optimization. *Wirel. Netw.* **2020**, *26*, 2983–2998. [CrossRef]
17. Boyineni, S.; Kavitha, K.; Sreenivasulu, M. Mobile sink-based data collection in event-driven wireless sensor networks using a modified ant colony optimization. *Phys. Commun.* **2022**, *52*, 101600. [CrossRef]
18. Donta, P.K.; Amgoth, T.; Annavarapu, C.S.R. An extended ACO-based mobile sink path determination in wireless sensor networks. *J. Ambient Intell. Hum. Comput.* **2021**, *12*, 8991–9006. [CrossRef]
19. Ghaleb, M.; Subramaniam, S.; Ghaleb, S.M. An Adaptive Data Gathering Algorithm for Minimum Travel Route Planning in WSNs Based on Rendezvous Points. *Symmetry* **2019**, *11*, 1326. [CrossRef]
20. Sun, G.; Liu, Y.; Zhang, J.; Wang, A.; Zhou, X. Node Selection Optimization for Collaborative Beamforming in Wireless Sensor Networks. *Ad Hoc Netw.* **2016**, *37*, 389–403. [CrossRef]
21. Hartigan, J.A.; Wong, M.A. Algorithm as 136: A k-means clustering algorithm. *Appl. Stat.* **1979**, *28*, 100–108. [CrossRef]
22. Zadeh, L.A. Fuzzy sets. *Inf. Control* **1965**, *8*, 338–353. [CrossRef]
23. Heinzelman, W.R.; Chandrakasan, A.; Balakrishnan, H. Energy-efficient communication protocol for wireless micro sensor networks. In Proceedings of the 33rd Annual Hawaii International Conference on System Sciences, Maui, HI, USA, 7 January 2000; pp. 1–10. [CrossRef]
24. Banimelhem, O.; Abu-hantash, A. Fuzzy logic-based clustering approach with mobile sink for WSNs. In Proceedings of the 13th International Computer Engineering Conference (ICENCO), Cairo, Egypt, 27–28 December 2017. [CrossRef]
25. Johnson, D.S.; McGeoch, L.A. Experimental analysis of heuristics for the STSP. In *The Traveling Salesman Problem and Its Variations*; Springer: Boston, MA, USA, 2007; pp. 369–443. [CrossRef]

computers

MDPI

Article

Energy-Efficient Cluster Head Selection in Wireless Sensor Networks Using an Improved Grey Wolf Optimization Algorithm

Mandli Rami Reddy [1,2], M. L. Ravi Chandra [2], P. Venkatramana [3] and Ravilla Dilli [4,*]

1 Electronics and Communication Engineering Department, Jawaharlal Nehru Technological University Ananthapuramu, Ananthapuramu 515002, Andhra Pradesh, India
2 Electronics and Communication Engineering Department, Srinivasa Ramanujan Institute of Technology, Ananthapuramu 515701, Andhra Pradesh, India
3 Electronics and Communication Engineering, Mohan Babu University (Sree Vidyanikethan Engineering College), Tirupati 517102, Andhra Pradesh, India
4 Electronics and Communication Engineering, Manipal Institute of Technology, Manipal Academy of Higher Education, Manipal, Udupi 576104, Karnataka, India
* Correspondence: dilli.ravilla@manipal.edu; Tel.: +91-9035190124

Abstract: The internet of things (IoT) and industrial IoT (IIoT) play a major role in today's world of intelligent networks, and they essentially use a wireless sensor network (WSN) as a perception layer to collect the intended data. This data is processed as information and send to cloud servers through a base station, the challenge here is the consumption of minimum energy for processing and communication. The dynamic formation of cluster heads and energy aware clustering schemes help in improving the lifetime of WSNs. In recent years, grey wolf optimization (GWO) became the most popular feature selection optimizing, swarm intelligent, and robust metaheuristics algorithm that gives competitive results with impressive characteristics. In spite of several studies in the literature to enhance the performance of the GWO algorithm, there is a need for further improvements in terms of feature selection, accuracy, and execution time. In this paper, we have proposed an energy-efficient cluster head selection using an improved version of the GWO (EECHIGWO) algorithm to alleviate the imbalance between exploitation and exploration, lack of population diversity, and premature convergence of the basic GWO algorithm. The primary goal of this paper is to enhance the energy efficiency, average throughput, network stability, and the network lifetime in WSNs with an optimal selection of cluster heads using the EECHIGWO algorithm. It considers sink distance, residual energy, cluster head balancing factor, and average intra-cluster distance as the parameters in selecting the cluster head. The proposed EECHIGWO-based clustering protocol has been tested in terms of the number of dead nodes, energy consumption, number of operating rounds, and the average throughput. The simulation results have confirmed the optimal selection of cluster heads with minimum energy consumption, resolved premature convergence, and enhanced the network lifetime by using minimum energy levels in WSNs. Using the proposed algorithm, there is an improvement in network stability of 169.29%, 19.03%, 253.73%, 307.89%, and 333.51% compared to the SSMOECHS, FGWSTERP, LEACH-PRO, HMGWO, and FIGWO protocols, respectively.

Keywords: IoT; IIoT; wireless sensor network; grey wolf optimizer; energy-efficient; improved grey wolf optimizer

Citation: Rami Reddy, M.; Ravi Chandra, M.L.; Venkatramana, P.; Dilli, R. Energy-Efficient Cluster Head Selection in Wireless Sensor Networks Using an Improved Grey Wolf Optimization Algorithm. *Computers* **2023**, *12*, 35. https://doi.org/10.3390/computers12020035

Academic Editor: Sergio Correia

Received: 16 December 2022
Revised: 26 January 2023
Accepted: 2 February 2023
Published: 6 February 2023

1. Introduction

WSNs became the backbone of all the smart IoT applications, and their reliable deployment is very important for diverse real-time applications like the military, industry, wide-area surveillance, environmental monitoring factors, and health monitoring. WSNs play an important role in the Industry 4.0 revolution and they are essential in the perception/sensing layer of IoT systems for sensing the physical environment and collecting the data using SNs. Due to the short span of battery life in the SNs of WSNs, optimal energy consumption has been always a challenge. The energy efficiency of sensor nodes plays a major role due to their constrained resources in terms of processing and communication. Therefore, it is essential to propose efficient energy consumption algorithms to extend the lifetime and stability of WSNs. Clustering is a prominent mechanism to achieve energy efficiency in WSNs. Clustering-based architecture in WSNs reduces the number of data transmissions using intra-cluster and inter-cluster communications [1]. However, the performance of clustering depends on the process of CH selection and the formation of optimal number of clusters. In a cluster-based architecture, a random selection of CHs causes poor connectivity, unexpected node failures, and reduces network lifetime. On the other hand, optimal selection of CHs enhances the performance and lifetime in WSNs. An optimized routing algorithm through an efficient CH selection process is essential for larger-scale WSNs. Clustering-based routing supports load balancing, reliable communication, and fault tolerance to prolong the life time of a WSN. CH selection based on node position, centrality of nodes, residual energy, number of neighbors, and node rank (rank is assigned depends on number of links, link cost) overcomes the drawbacks of the LEACH protocol [2]. Dynamic and on-demand CH selection based on the occurrence of events minimizes the message and computational overhead, and also ensures energy balancing among CHs [3]. Optimal clustering replaces one-hop communication between the CHs and sink node by an optimal multi-hop distance in order to mitigate energy consumption and enhance the network lifetime by 35% in WSNs [4]. Adaptive CH selection in heterogeneous WSNs, based on residual energy and node location, ensures that the node which has higher residual energy and is close to the BS becomes the CH with the highest probability [5].

In recent years, GWO became the most popular feature selection optimization, swarm intelligent, and robust metaheuristics algorithm that gives competitive results in solving engineering problems. In spite of several studies in the literature to enhance the performance of the GWO algorithm, there is a need for further improvements in terms of the balance between exploitation and exploration, lack of population diversity, and premature convergence of the basic GWO algorithm. In this paper, we have proposed an energy-efficient CH selection using an improved version of the GWO (EECHIGWO) algorithm to enhance the energy efficiency, average throughput, network stability, and the network lifetime in WSNs with an optimal selection of CHs. It considers sink distance, residual energy, CH balancing factor, and average intra-cluster distance as the parameters in selecting the CH. The simulation results have confirmed the optimal selection of the CH with minimum energy consumption, resolved premature convergence, and enhanced the network lifetime by using minimum energy levels in WSNs.

The definition of fitness functions plays a key role in selecting optimal CHs in WSNs. In the existing literature, the fitness functions are defined with equal or random weight values irrespective of the SN's position and its available energy. The novelty towards the proposed work include the computation of optimal fitness value for a given SN based on the residual energy and its distance to BS. For optimal clustering and routing, objective functions are considered where routing fitness is computed based on the minimum number of hops, mean load, and distance between gateways and BS. In this paper, the minimum value of all fitness functions of gateways is considered as clustering fitness function.

The remainder paper is organized as follows: the main contributions and literature survey of this work are described in Sections 2 and 3, respectively. The clustering based on the proposed EECHIGWO algorithm is explained in Section 4. Section 5 explains the results and discussions of the proposed method and comparison with the recently proposed GWO-based CH selection methods. Finally, the conclusions and future scope are made in Section 6.

2. Contributions

Motivation

Due to the limited resources of SNs in WSNs and applications where recharging or replacing the battery is not a feasible solution, it is essential to design and implement energy-efficient schemes to improve the key performance parameters. Even though clustering is considered to be the most prominent technique to prolong the lifetime expectancy in WSNs, the process of CH selection in order to enhance the network lifetime is still a challenge. The conventional clustering-based routing algorithms support fault tolerance, load balancing, and reliable communications at the cost of decreased lifetime of the CH. To overcome this, there has been a continuous research on designing efficient CH selection techniques, data acquisition, and routing optimization algorithms.

In this article, an improved version of the GWO algorithm is applied for an optimal CH selection in WSNs to minimize the energy levels used for computation and communication. The performance of the proposed protocol is evaluated in terms of the number of dead nodes, energy consumption levels, the number of operating rounds, and the average throughput. A rigorous statistical analysis and simulations are carried out by taking the average of fifteen readings for each result to prove the proposed algorithm's efficiency. In addition, a comparative analysis is performed with the recently proposed GWO-based algorithms. The simulation results prove that the proposed algorithm outperforms in terms of energy preservation and an enhanced network lifetime.

The main contributions of this article are as follows:

(a) Rigorous literature study of algorithms and protocols are conducted that enhances the WSN lifetime with an optimal CH selection and energy-efficient techniques.
(b) Study of the futuristic algorithms proposed based on the GWO algorithm for CH selection and optimal energy utilization in WSNs.
(c) Proposed a novel method based on an improved GWO algorithm, distance between BS and SN for CH selection, and efficient energy utilization.
(d) Defined the fitness function based on the IGWO algorithm that considers residual energy at SN to avoid randomness in CH selection for energy-efficient data deliveries.
(e) Compared the performance of the proposed algorithm with existing GWO-based algorithms in terms of the number of dead nodes, number of operating rounds, energy consumption, and the average throughput.
(f) Proved that the proposed EECHIGWO algorithm outperforms the existing GWO-based algorithms in WSNs.

3. Literature Survey

Efficient energy utilization is one of the primary goals to maximize the network lifetime in WSNs. Clustering is known to be one of the efficient techniques in WSNs to enhance energy efficiency by designing an energy-efficient protocol in CHs selection. There are various techniques present in the literature for electing CHs in WSNs to enhance the network lifetime, but this still remains a major challenge in WSNs.

3.1. *Energy Efficient Techniques for WSNs*

The energy efficiency is a critical parameter to be addressed in WSNs, as the individual SNs operate with limited energy sources and optimizing the energy consumption of SNs has been a challenging design issue in WSNs. Energy-efficient WSNs compromise with network stability as a crucial factor in ensuring long lasting and reliable network coverage. Clustering and routing are essential aspects to be considered for the efficient energy consumption of SNs in WSNs [6]. An adaptive hierarchical routing and hybrid clustering based on the fuzzy C-means method, residual energy, BS location, and Euclidean distance improves coverage and lifetime of the network [7]. Fuzzy-based clustering provides energy-efficient routing capabilities that enhances the network lifetime [8].

Equalized CH election routing ensures the energy conservation in a balanced fashion and enhances the network lifetime [9]. A neuro-fuzzy-based energy-aware clustering is proposed in WSNs that consist of neural networks and a fuzzy subsystem to achieve energy-efficient clusters and CHs. The performance of these systems are measured based on residual energy, transmission range, and trust factor (for security) [10]. Multi-level route-aware clustering minimizes routing control packets and moderates the energy consumption at relay nodes present near the BS [11]. Formation of clusters in WSNs based on the Voronoi diagram minimizes the energy consumption for communication. This method can enhance the FND by 14.5% compared to SEP [12].

3.2. Energy Aware Clustering and Performance Optimization Using Metahueristic Approach

In this section, the importance of metaheuristic algorithms to solve the engineering problems and the role of GWO in enhancing the performance of WSNs are highlighted. The energy constraints in measuring network lifetime pose a challenge in a widely spread applications of WSNs. Network stability and energy efficiency are two typical trade-off parameters in WSNs. There have been continuous efforts by researchers in achieving the energy efficiency in WSNs that includes state-of-the-art metaheuristic algorithms [13].

In recent years, swarm intelligence metaheuristic optimization techniques have proved the outstanding performers in solving a wide range of engineering and science problems. GWO is one such technique, and it became popular due to the involvement of only few parameters and no derivation information. It provides right balance between exploitation and exploration that leads to favorable convergence. It has applications in the fields of networking, image processing, machine learning, bioinformatics, global optimization, environmental applications, etc. [14]. For enhancing the efficient usage of computational resources, an adaptive GWO tunes the exploitation and exploration parameters automatically based on fitness function, and this reduces the number of iterations needed [15].

There have been many energy-efficient clustering protocols based on the GWO algorithm proposed in the recent times towards optimal CH selection [16,17]. GWO-based methods are proposed for energy optimizations in WSNs by finding the optimal positions of SNs to achieve maximum connectivity and coverage. It has been shown that the GWO-based CH selection algorithm performs better than PSO, GA, and Greedy approaches [18,19]. Precision improvement of the SN positions improves the data transmission among SNs in the network, saves the node's energy, and also enhances the network lifetime [20].

The GWO algorithm is used to define a connected dominating set based on distance and it is used to achieve energy efficiency and stability in cluster-based WSNs [21]. A GWO algorithm-based approach enhances the energy efficiency compared to ABC and AFS algorithms [22]. The GWO-based game theoretical approach gives better solutions in selecting optimal aggregation points to improve the SN's battery lifetime [23]. The SMO algorithm is proposed based on the sampling population for energy efficient CH selection [24]. Multi-object-based SMO is an energy efficient clustering and routing algorithm that balances the load at gateways for an improved network lifetime compared to PSO and GWO algorithms [25]. A combination of using GWO and whale optimization algorithms for clustering and dynamic CH selection increases the capabilities in terms of exploitation and exploration [26]. A whale optimization-based algorithm improves the rate of utilization of SNs and coverage in heterogeneous WSNs [27]. To reduce the energy consumption in CH selection, an objective function in GWO is defined based on residual energy, intra-cluster distance, CH balancing factor, and sink distance [28]. Topology control based on binary GWO introduces a fitness function that reduces the number of active nodes and enhances the network lifetime [29].

In the literature, the state-of-the-art metaheuristic algorithms like ACO, BA, GA, PSO, WOA, MFO, etc., are proposed to solve the optimization problems in engineering applications. COA integrated with a dimension learning-based hunting strategy maintains diversity and enhances the balance between exploration and exploitation. It effectively provides the optimization of energy constraints in WSNs [30]. CSA is used to select the optimal CHs in heterogeneous WSNs in order to improve the energy efficiency and network lifetime compared to PSO, GA, and LEACH algorithms [31]. A BSO swarm-based algorithm helps in selecting best possible CHs for enhancing the coverage, data rate, and energy efficiency in WSNs [32]. An ARSH-FATI-based CH selection dynamically (at runtime) switches between exploitation and exploration of the search process. It enhances the network lifetime by 25% compared to a PSO algorithm [33]. Network coverage optimization in heterogeneous WSNs using a sine-cosine-based WOA balances the local and global search capabilities, speeds up the search process, and enhances the optimization accuracy. It also maximizes the utilization rate of nodes, thereby mitigating the network cost.

Out of all the existing well-known metaheuristic algorithms (such as PSO, GSA, DE, EP, and ES), GWO has proved to be a powerful swarm-intelligent algorithm introduced to handle continuous and discrete optimization problems in the field of engineering. It is a unique metaheuristic algorithm that mimics the leadership hierarchy and attacking strategy of grey wolves. It is used for solving classic, real engineering design problems in unknown search spaces [34]. It improves the deterministic approach of a stochastic optimization for multi-robot exploration in the given space [35].

The GWO algorithm can be applied effectively in various fields of engineering and has many applications. It is applied in the image processing domain that includes image segmentation, image compression, image classification, and medical imaging to enhance efficiency and robustness [36]. GWO is used to enhance the accuracy of the IDS by 81% in detecting anomalous traffic in the network [37–39]. It improves the task allocation process and minimizes runtimes of the serverless frameworks for cloud applications at varied load conditions [40]. It is used for the secure transfer of data in IoT applications in which the GWO-based security algorithms offer lower memory and time for encryption/decryption [41]. GWO is useful for text clustering in text mining application to improve convergence rate and avoid trapping into local minima. The combination of GWO and GO algorithms give 87.6% efficiency in terms of precision, accuracy, recall, and sensitivity compared to individual algorithms [42].

3.3. Role of GWO Algorithm in Optimal CH Selection

The optimal CH selection using GWO greatly enhances network performance in terms of coverage, throughput, energy consumption, and network lifetime in WSNs [43]. It formulates the objective function and its weights based on intra-cluster distance, CH balancing factor, residual energy, and sink distance [44–46]. GWO addresses clustering and routing issues by formulating an optimal fitness function so that the number of hops and overall distance traversed are minimized, and also load balancing is achieved. The fitness functions for routing and clustering give higher values compared to GA and PSO algorithms [47]. A hybrid approach of GWO and WOA provides effective cluster formation, dynamic CH selection, and an optimal number of CHs in WSNs. It has better exploration and exploitation capabilities than the individual optimization approaches. CH selection based on the combination of GWO and CSO algorithms avoids premature convergence in exploring the search space. It gives a trade-off between the exploration and exploitation in CH selection to enhance the network lifetime expectancy more than FFO, ABCO, FGGWO algorithms [48].

Distance-based stable CDS along with GWO provides an enhanced performance of 70.5% over the GA-based algorithms in terms of energy efficiency and network stability. A three-level hybrid clustering is proposed for WSNs using the GWO algorithm. At level 1, BS selects the CHs; in level 2, there is GWO-based optimal data routing; and in level 3, distributed clustering takes place. This hybrid clustering enhances the network performance in terms of residual energy, stability, and lifetime [49]. The network coverage optimization using a minimal distribution of redundant nodes can enhance the stability and lifetime in WSNs. The GWO algorithm embedded with SA can achieve better coverage optimization than PSO in terms of optimization speed, energy consumption, and network lifetime [50]. The coverage optimization in WSNs using a Virtual Force Levy-embedded GWO algorithm performs better than CSA and Chaotic PSO techniques in terms of scalability, adaptability, uniformity, and coverage rate [51].

GWO is used to compute the threshold levels of sensor decision rules at the fusion center without depending on initial values and provides lower complexity in WSNs [52]. It is used to localize the SNs with minimal position errors, and with a quicker convergence than PSO and MBA algorithms [53]. The quantum computing with a clone operation in the GWO algorithm avoids falling it into a local optimal solution. The optimal design of the sensor duty cycle in industrial WSNs using the quantum clone GWO improves convergence speed and network lifetime compared to GA and SA algorithms [54].

3.4. Enhanced Versions of GWO Algorithms for WSNs

The conventional GWO algorithm may give sub-optimal/local optimal solutions because of its minimal exploration at early stages. An improved GWO aims to enhance the optimization accuracy, accelerating the convergence of the GWO algorithm, and balancing between exploration and exploitation. There are various attempts that have been made to address the limitations of the GWO algorithm in terms of convergence speed, convergence accuracy, and instability [55]. The improved versions of the GWO algorithm are applied to WSNs for enhancing the convergence speed and precision. The features of these algorithms are presented in Table 1, and they attempted to overcome the issues of slow convergence, falling into local minima, and low search precision of the GWO algorithm [56].

Dimension-based learning in GWO addresses the drawbacks of the conventional GWO algorithm, i.e., a lack of population diversity, premature convergence, and the imbalance between exploration and exploitation. It demonstrates applicability and efficiency in solving engineering design problems in a superior way compared to the conventional GWO algorithm [57]. Weighted GWO enhances the convergence rate with higher exploration and exploitation in the searching space. A weighted GWO algorithm with an MLP neural network further enhances the classification accuracy with optimal weights [58]. Binary GWO with SVM is used to improve the intrusion detection rate and accuracy in WSNs. It improves the intrusion detection rate and detection accuracy, and at the same time it minimizes the processing time, number of features, and false alarm rates [59]. "Differential evolution" is introduced to update the wolf pack at each iteration based on the fitness value. 'R' wolves with the least fitness values are eliminated and new set of (randomly generated) wolves are introduced. It gives better optimization accuracy and convergence speed than CSA, PSO, and ABCO algorithms [60]. An improved version of the GWO algorithm supports an energy-efficient, balanced CH structure in WSNs based on fitness values, and it extends the network stability period and throughput by 31.5% compared to the SEP algorithm. Behavior-based GWO performs better in terms of population diversity and convergence. The objective function for this algorithm is defined by considering the connectivity rate, coverage rate, and total energy consumption in WSNs [61]. A fuzzy-extended, GWO algorithm-based, threshold-sensitive, energy-efficient clustering protocol enhances the network stability. A modified GWO algorithm for heterogeneous WSNs selects the initial clusters depending on the values of the fitness functions for energy nodes. The fitness values are considered as initial weights in GWO and these weights are updated dynamically based on the distance between the wolves and their prey. It ensures the selection of optimal CHs and enhances the network lifetime by 55.7% and 31.9%, compared to SEP and the distributed energy-efficient clustering algorithm, respectively [62].

Table 1. Comparison of the relevant algorithms and their features.

Protocol	Nodes Type	Inter-Cluster Topology	Need of Energy Awareness	CH Selection	Heuristic Approach
SSMOECHS [24]	Homogeneous	Single-hop	No	Probabilistic	No
GWO-C [43]	Homogeneous	Single-hop	No	Probabilistic	Yes
GWO-based clustering [44]	Homogeneous	Dual-hop	No	Probabilistic	Yes

Table 1. *Cont.*

Protocol	Nodes Type	Inter-Cluster Topology	Need of Energy Awareness	CH Selection	Heuristic Approach
GWO [47]	Heterogeneous	Multi-hop	Yes	Probabilistic	Yes
HGWCSOA [48]	Homogeneous	Single-hop	Yes	Probabilistic	Yes
QCGWO [54]	Homogeneous	Not applicable	No	Not applicable	Yes
BGWO [61]	Homogeneous	Single-hop	No	Probabilistic	Yes
FGWSTERP [62]	Homogeneous	Single-hop	Yes	Fuzzy based	Yes
LEACH-PRO [63]	Homogeneous	Single-hop	Yes	Probabilistic	No
HMGWO [64]	Heterogeneous	Single-hop	Yes	Probabilistic	Yes
FIGWO [65]	Homogeneous	Single-hop	Yes	Deterministic	Yes

4. Methodology

In this section, the proposed EECHIGWO algorithm is presented with details to enhance the network lifetime using an optimal CH selection process. This network model is meant mainly for industrial applications where the different manufacturing segments of a plant are located at different geographical places and the assumptions are as follows:

1. The SNs are randomly deployed in a two-dimensional geographical space.
2. The BS is located at the center of the network terrain and there is multi-hop communication from CHs to the BS.
3. The SNs are divided into approximately equal groups, and they are randomly distributed within the group.
4. The SNs are homogeneous within the group and their mobility is limited to 0.2 m/s.
5. BS and the nodes who participate in multi-path communication only will have uninterrupted power supply.
6. BS executes the algorithm for CH selection and also it collects the aggregated data from all CHs.

Figure 1 shows the radio energy model of an SN using two different channel models: free space path loss (d^2) model for a single-hop communication and multipath propagation fading (d^4) model for the multi-hop path communication. Therefore, the energy consumption for transmitting an n-bit packet over distance 'd' is computed as

$$E_{TX}(n, d) = \begin{cases} nE_{elec} + n\, e_{fs}\, d^2 & d < d_0 \\ nE_{elec} + n\, e_{mp}\, d^4 & d \geq d_0 \end{cases} \quad (1)$$

where

$e_{fs} \rightarrow$ energy dissipation coefficient of free-space attenuation model
n $\rightarrow$ packet length
$e_{mp} \rightarrow$ energy dissipation coefficient of multipath attenuation model
d $\rightarrow$ distance between sender and receiving node
$d_0 = \sqrt{e_{fs} / e_{mp}} \rightarrow$ threshold distance
$E_{elec} \rightarrow$ energy needed to transmit/receive 1-bit data.

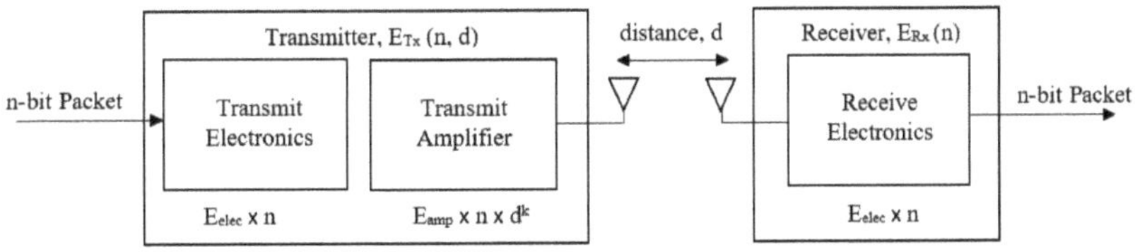

Figure 1. Radio Energy Model of a Sensor Node [62].

At Rx, the amount of energy consumption for receiving n-bit data packet is computed as

$$E_{RX}(n) = n \times E_{elec} \tag{2}$$

There are three parameters that contribute to the energy consumption at CH: number of data packets received from SNs which are members of a particular cluster, data aggregation performed by CH, and number of aggregated packets transmitted from CH to BS. Therefore, the energy consumption at CH is given as

$$E_{CH} = E_{RX}(n,\ d) \times SN_{num} + E_{DF} \times\ n\ \times (SN_{num} + 1) + E_{TX}(n,\ d) \tag{3}$$

$SN_{num} \rightarrow$ SN's number in a particular cluster $E_{DF} \rightarrow$ data fusion energy/bit.

For all SNs other than CHs, the energy consumption is E_{TX} (n, d).

The total remaining energy during the kth round is computed as:

$$E_R(k) = E_R(k-1) - \left(\sum_{l=1}^{CH_{num}(l)} E_{CH}(l) + \sum_{m=1}^{SN_{alive}(k)-CH_{num}(k)} E_{SN}(m) \right) \tag{4}$$

$E_R(k-1) \rightarrow$ total remaining energy at $(k-1)^{th}$ round
$CH_{num}(k) \rightarrow$ number of CHs in the kth round
SN_{alive} (k) $\rightarrow$ total number of alive nodes in the kth round
$E_{CH}(l) \rightarrow$ energy consumed by lth CH
$E_{CH}(m) \rightarrow$ energy consumed by mth SN

Proposed EECHIGWO Algorithm

To overcome the randomness in CH selection, BS performs CH selection based on the proposed EECHIGWO algorithm. The information about the selected CHs are broadcasted to all SNs through the multi-hop communication nodes. The total number of SNs are divided into four subsets (approximately) based on fitness value and out of which the location of sixteen SNs are declared as fixed to support multi-hop paths. The SNs are considered as grey wolves and CH is the prey. The EECHIGWO algorithm is defined in terms of rounds, and each round consists of CH formation stage and data transmission stage. The fitness value of an SN is computed based on the residual energy and its distance to BS.

$$F = \begin{cases} 0.8 \times \left(\frac{E_{residual}}{E_{initial}}\right) + 0.2 \times \left(\frac{d_{max} - d}{d_{max} - d_{min}}\right), & E_{residual} < 0 \\ 0.2 \times \left(\frac{E_{residual}}{E_{initial}}\right) + 0.8 \times \left(\frac{d_{max} - d}{d_{max} - d_{min}}\right), & E_{residual} \geq d_0 \end{cases} \tag{5}$$

where

$E_{inital} \rightarrow$ initial energy of SN,
$E_{residual} \rightarrow$ residual energy at SN after each round,
d $\rightarrow$ distance between SN and BS,
$d_{max} \rightarrow$ maximum distance between SN and BS,
$d_{min} \rightarrow$ minimum distance between SN and BS.

The Equation (5) represents the fitness function in which 80% weightage is given to residual energy at SN and 20% weightage is given to the distance between the SN and BS.

The initial position of BS is computed as

$$\overrightarrow{X_{CH}} = \left| \omega_\alpha \overrightarrow{X_\alpha} + \omega_\beta \overrightarrow{X_\beta} + \omega_\delta \overrightarrow{X_\alpha} \right| \tag{6}$$

where ω_α, ω_β, ω_δ are the initial weights and they are calculated as follows:

$$\omega_\alpha = \frac{F_\alpha}{F_\alpha + F_\beta + F_\delta}, \ \omega_\beta = \frac{F_\beta}{F_\alpha + F_\beta + F_\delta}, \ \omega_\delta = \frac{F_\delta}{F_\alpha + F_\beta + F_\delta} \tag{7}$$

F_α, F_β, F_δ are the first three optimal fitness values of the SNs.

To enhance the capabilities of global search using the GWO algorithm, the weights ω_α, ω_β, ω_δ are dynamically updated using the vectors $\vec{D}$, $\vec{A}$ and at the ith iteration, the weights are calculated as:

$$\omega_\alpha^{i+1} = \frac{\overrightarrow{D_\alpha^{i+1}} \times \overrightarrow{A_\alpha^{i+1}}}{\overrightarrow{D_\alpha^{i+1}} \times \overrightarrow{A_\alpha^{i+1}} + \overrightarrow{D_\beta^{i+1}} \times \overrightarrow{A_\beta^{i+1}} + \overrightarrow{D_\delta^{i+1}} \times \overrightarrow{A_\delta^{i+1}}} \tag{8a}$$

$$\omega_\beta^{i+1} = \frac{\overrightarrow{D_\beta^{i+1}} \times \overrightarrow{A_\beta^{i+1}}}{\overrightarrow{D_\alpha^{i+1}} \times \overrightarrow{A_\alpha^{i+1}} + \overrightarrow{D_\beta^{i+1}} \times \overrightarrow{A_\beta^{i+1}} + \overrightarrow{D_\delta^{i+1}} \times \overrightarrow{A_\delta^{i+1}}} \tag{8b}$$

$$\omega_\delta^{i+1} = \frac{\overrightarrow{D_\delta^{i+1}} \times \overrightarrow{A_\delta^{i+1}}}{\overrightarrow{D_\alpha^{i+1}} \times \overrightarrow{A_\alpha^{i+1}} + \overrightarrow{D_\beta^{i+1}} \times \overrightarrow{A_\beta^{i+1}} + \overrightarrow{D_\delta^{i+1}} \times \overrightarrow{A_\delta^{i+1}}} \tag{8c}$$

During the CH selection process, the location of the CH is computed using α, β, ω wolves, and the other SNs compute their distances with respect to BS as shown in Figure 2.

The updated position of SN in the $(i + 1)^{th}$ iteration is computed as:

$$\overrightarrow{X^{i+1}} = \overrightarrow{X_{CH}^{i}} - \vec{A} \times \vec{D} \tag{9}$$

where $\vec{A}$ is the convergence vector and it is given as $\vec{A} = 2\vec{\alpha} \times \vec{r_1} - \vec{\alpha}$, $\overrightarrow{X_{CH}^{i}}$ is the CH position in the previous iteration, i.e., i^{th} iteration.

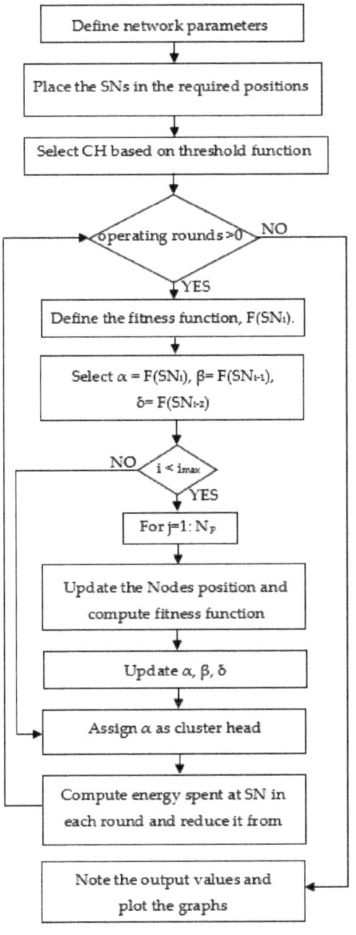

Figure 2. Flow diagram of the proposed EECHIGWO Algorithm.

5. Results and Discussions

The performance of the proposed EECHIGWO algorithm is evaluated by conducting extensive simulations in MATLAB 2022B. Each simulation reading is considered by taking an average of fifteen simulation runs and the results are compared with the existing state-of-the-art literature in enhancing the energy efficiency of WSNs using GWO-based techniques. With the same experimental parameters as shown in Table 2, the EECHIGWO algorithm's performance is compared with the SSMOECHS [24], FGWSTERP [62], LEACH-PRO [63], HMGWO [64], and FIGWO [65] algorithms. For ease of comparison, the number of SNs are considered as 100 in the network terrain of 100 m^2. The metrics considered for analyzing the performance of the proposed algorithm include average energy consumption, number of dead nodes to define the network stability, and average throughput which defines the number of data packets delivered to BS. During the operation of WSNs, the SNs send the sensed information to their respective CHs and each CH forwards the aggregated information from various SNs to BS through fixed intermediate nodes.

Table 2. Initial parameters of EECHIGWO protocol for simulations.

Parameter	Value
Network Terrain	100 m^2
Network size	100
Initial Energy (E_0)	1 J
Probability to become CH (P)	0.1
Number of CHs	P × 100
E_{fs}, E_{elec}, E_{amp}	10 pJ/bit/m^2, 50 nJ/bit, 0.0013 pJ/bit/m^4
$D_{critical}$, D_{max}	20 m, 100 m
Data Packet size	500-Bytes
BS Position	(50, 50)

Network lifetime can be defined based on stable and unstable periods. The stable period is the time at which network starts operating till the FND. The unstable period is the time duration between the FND and LND. In Figure 3, the FND, HND, and LND are observed at 5940th, 6604th, 7908th operating rounds, respectively. The reason for this enhanced lifetime of the proposed EECHIGWO protocol is that it eliminates the random selection of CH. The BS selects the CHs based on the optimal fitness values of the SNs, and the CHs' selection information is broadcasted to all SNs through only the multi-hop communication nodes, which are placed in the fixed locations with an uninterrupted power supply. Therefore, the SNs with lower residual energy levels have a lower probability of being elected as CH. It enhances the network lifetime by avoiding the sudden death of SNs who have lower residual energies.

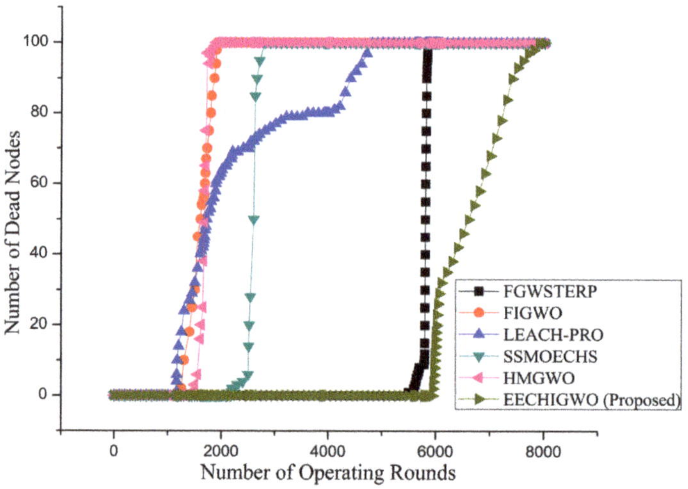

Figure 3. Network lifetime during stable and unstable periods.

The global optimization capabilities of the proposed protocol give balanced energy consumption among SNs and leads to minimal average energy consumption at each round, as shown in Figure 4. The energy consumption is minimized due to optimal intra-cluster communication, uniform distribution of clusters, and multi-hop routing based on the distance between CH and BS. The multi-hop communication feature of the proposed algorithm enhances the load balancing capabilities and mitigates the energy consumption at CHs located far away from BS. From Figure 4, it can be seen that the first SN's death was much later than the other protocols. Similarly, in the given network size of 100 SNs, the 50% SNs death, 100% SNs death was significantly increased compared to other protocols. The overall observation is that the proposed EECHIGWO protocol gives superior performance in terms of network lifetime compared to other protocols shown in Figure 4.

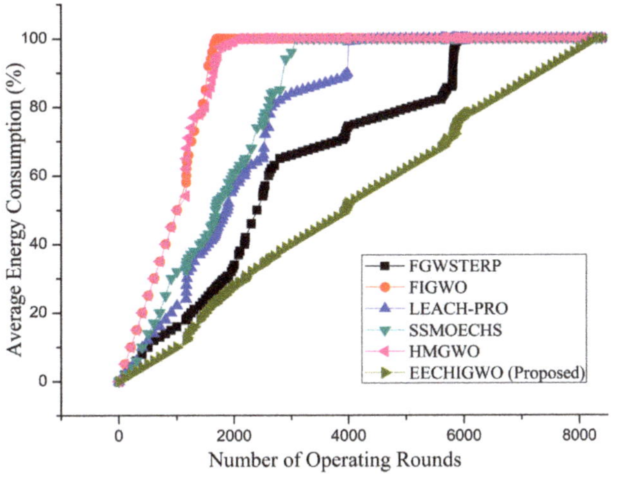

Figure 4. The average energy consumption of EECHIGWO algorithm compared with other protocols.

The throughput is measured in terms of the number of data packets delivered to BS from the SNs. The proposed EECHIGWO protocol provides higher throughput than other protocols as shown in Figure 5. This is due to the fact that it adopts an optimal CH selection and the SNs have the highest survival time. The even distribution of energy consumption at SNs improves the network throughput as well as prolongs the network life. At the given number of rounds, the number of alive SNs in the network are higher than that of the other algorithms; therefore, more data groups are generated, and the number of data packets delivered at the BS also increases. At a higher number of operating rounds, particularly after the round number 1600, the throughput using EECHIGWO is much higher than other protocols where more data packets are generated towards the BS.

Table 3 shows the network stability in terms of FND, HND, and LND of the proposed protocol and compares with various existing protocols. From the readings shown in Table 3, the rapid death of SNs is reduced from the round where FND occurs using the proposed algorithm. This is because of the criteria that the SN with lower residual energies become CH with very minimal probability. There is an improvement in network stability of 169.29%, 19.03%, 253.73%, 307.89%, and 333.51% compared to the SSMOECHS, FGWSTERP, LEACH-PRO, HMGWO, and FIGWO protocols, respectively.

Table 3. Network stability comparison in terms of number of rounds.

Algorithm	FND	FND Improvement (%)	HND	HND Improvement (%)	LND	LND Improvement (%)	Overall Improvement (%)
SSMOECHS [24]	2190	171.23	2600	154	2798	182.63	169.29
FGWSTERP [62]	5500	8	5807	13.72	5841	35.38	19.03
LEACH-PRO [63]	1159	412.5	1720	283.95	4800	64.75	253.73
HMGWO [64]	1450	309.65	1675	294.27	1884	319.75	307.89
FIGWO [65]	1248	375.96	1612	309.68	1906	314.9	333.51
EECHIGWO [Proposed]	5940	—	6604	—	7908	—	—

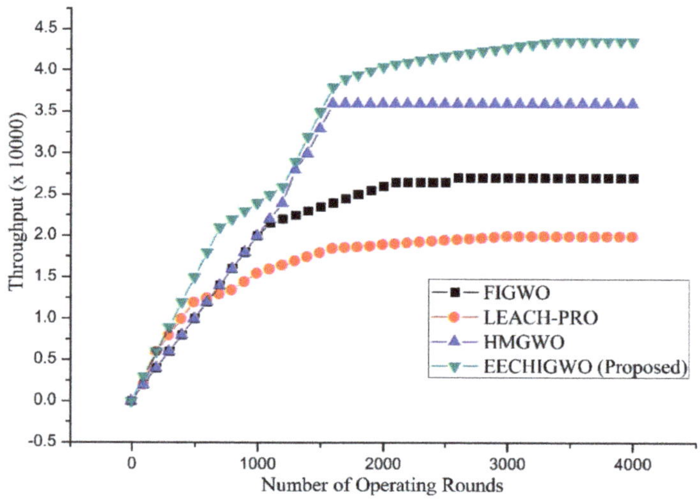

Figure 5. Average throughput of EECHIGWO protocol comparison with other protocols.

6. Conclusions

In this paper, an energy-efficient CH selection using an improved version of the GWO algorithm is proposed which considers sink distance, residual energy, balancing factor, and average intra-cluster distance as the parameters in selecting the CH. The proposed EECHIGWO protocol has multi-hop features and provides optimal fitness function values to improve the WSN's lifetime. The design of fitness function for CH selection is based on both the amount of residual energy at SNs and their Euclidean distance to BS. It supports deterministic and even selection of CHs in each round that leads to balanced energy consumption and avoids premature deaths of SNs. The performance of the protocol is tested in terms of number of dead nodes to define the network stability, average energy consumption, number of operating rounds, average throughput, and network lifetime. The simulation results have confirmed the optimal selection of CH with minimum energy consumption. It is proved that the network throughput, stability, and the network lifetime are enhanced compared to the existing state-of-the-art energy-efficient routing protocols for WSNs such as FGWSTERP [62], FIGWO [65], LEACH-PRO [63], SSMOECHS [24], and HMGWO [64], which are single-hop protocols with higher energy consumption and provide lower network lifetime. Using the proposed algorithm, there is an improvement in network stability of 169.29%, 19.03%, 253.73%, 307.89%, and 333.51% compared to the SSMOECHS, FGWSTERP, LEACH-PRO, HMGWO, and FIGWO protocols, respectively. As a future scope of the current research, the performance of the proposed algorithm can be tested for heterogeneous WSNs with larger number of SNs and higher node densities.

Author Contributions: Conceptualization, R.D. and M.R.R.; methodology, R.D.; software, M.L.R.C.; validation, R.D., P.V.; and M.R.R.; formal analysis, M.L.R.C.; investigation, R.D., P.V.; resources, M.L.R.C.; data curation, M.R.R.; writing—original draft preparation, R.D.; writing—review and editing, M.L.R.C.; visualization, P.V.; supervision, M.L.R.C. All authors have read and agreed to the published version of the manuscript.

Funding: This research received no external funding.

Institutional Review Board Statement: Not applicable.

Informed Consent Statement: Not applicable.

Data Availability Statement: Not applicable.

Conflicts of Interest: The authors declare no conflict of interest.

Glossary

ABC	artificial bee colony optimization
ACO	ant colony optimization
AFS	artificial Fish Schooling
BA	bat algorithm
BGWO	behavior-based grey wolf optimizer
BS	base station
CDMA	code division multiple access
CDS	connected dominating set
CH	cluster head
COA	chimp optimizer algorithm
CSA	cuckoo search algorithm
CSO	crow search optimization
DEEC	distributed energy efficient clustering
DE	differential evolution
DLH	dimension learning-based hunting
EP	evolutionary programming
ES	evolution strategy
FCGWO	firefly cyclic grey wolf optimization
FFO	firefly optimization
FGWSTERP	fuzzy GWO based stable threshold sensitive energy efficient cluster based routing protocol
FIGWO	fitness value based Improved GWO
FND	first node death
GA	genetic algorithm
GOA	grasshopper optimization algorithm
GSA	gravitational search algorithm
GWO	grey wolf optimization
GWO-C	GWO with clustering
HGWCSOA	hybrid grey wolf and crow search optimization algorithm
HMGWO	modified GWO for heterogeneous WSN
HND	half node death
HWGWO	hybrid whale and grey wolf optimization
IDS	intrusion detection system
IGWO	improved grey wolf optimization
IIoT	industrial IoT
IoE	internet of everything
IoT	internet of things
LEACH	low-energy adaptive clustering hierarchy
LND	last node death
MBA	modified bat algorithm
MFO	moth-flame optimization
MLHP	multilayer hierarchical routing protocol
MLP	multi-layer perceptron
PRO	probabilistic cluster head selection
PSO	particle Swarm Optimization
QCGWO	quantum clone grey wolf optimization
SA	simulated annealing
SEP	stable election protocol
SMO	spider Monkey Optimization
SN	sensor node
SSMOECHS	sampling based spider monkey optimization and energy efficient cluster head selection
WOA	whale optimization algorithm
WSN	wireless sensor network

References

1. Al-Baz, A.; El-Sayed, A. A new Algorithm for Cluster Head Selection in LEACH protocol for Wireless Sensor Networks. *Int. J. Commun. Syst.* **2018**, *31*, e3407. [CrossRef]
2. Pour, S.E.; Javidan, R. A new energy aware cluster head selection for LEACH in wireless sensor networks. *IET Wirel. Sens. Syst.* **2021**, *11*, 45–53. [CrossRef]
3. Ghosal, A.; Halder, S.; Das, S.K. Distributed on-demand clustering algorithm for lifetime optimization in wireless sensor networks. *J. Parallel Distrib. Comput.* **2020**, *141*, 129–142. [CrossRef]
4. Arghavani, M.; Esmaeili, M.; Maryam, E.; Mohseni, F.; Arghavani, A. Optimal energy aware clustering in circular wireless sensor networks. *Ad Hoc Netw.* **2017**, *65*, 91–98. [CrossRef]
5. Zhongdong, H.; Hualin, W.; Zhendong, W. Energy balanced adaptive clustering routing protocol for heterogeneous wireless sensor networks. *Int. J. Wirel. Mob. Comput.* **2019**, *16*, 264–271. [CrossRef]
6. Khedr, A.M.; Pravija, R.P.V.; Ali, A.A. An Energy-Efficient Data Acquisition Technique for Hierarchical Cluster-Based Wireless Sensor Networks. *J. Wirel. Mob. Netw. Ubiquitous Comput. Dependable Appl.* **2020**, *11*, 70–86. [CrossRef]
7. Akhilesh, P.; Rajat, K.S. EEHCHR: Energy Efficient Hybrid Clustering and Hierarchical Routing for Wireless Sensor Networks. *Ad Hoc Netw.* **2021**, *123*, 102692. [CrossRef]
8. Raj, J.S.; Basar, A. QoS optimization of energy efficient routing in IoT wireless sensor networks. *J. ISMAC* **2019**, *1*, 12–23. [CrossRef]
9. Nikolidakis, S.A.; Kandris, D.; Vergados, D.D.; Douligeris, C. Energy Efficient Routing in Wireless Sensor Networks Through Balanced Clustering. *Algorithms* **2013**, *6*, 29–42. [CrossRef]
10. Robinson, Y.H.; Julie, E.G.; Balaji, S.; Ayyasamy, A. Energy Aware Clustering Scheme in Wireless Sensor Network Using Neuro-Fuzzy Approach. *Wirel. Pers. Commun.* **2017**, *95*, 703–721. [CrossRef]
11. Sabet, M.; Naji, H. An energy efficient multi-level route-aware clustering algorithm for wireless sensor networks: A self-organized approach. *Comput. Electr. Eng.* **2016**, *56*, 399–417. [CrossRef]
12. Liang, H.; Yang, S.; Li, L.; Gao, J. Research on routing optimization of WSNs based on improved LEACH protocol. *EURASIP J. Wirel. Commun. Netw.* **2019**, *194*, 194. [CrossRef]
13. Manshahia, M.S. Grey Wolf Algorithm based Energy-Efficient Data Transmission in Internet of Things. *Proc. Comput. Sci.* **2019**, *160*, 604–609. [CrossRef]
14. Faris, H.; Aljarah, I.; Al-Betar, M.A. Grey wolf optimizer: A review of recent variants and applications. *Neural Comput. Appl.* **2018**, *30*, 413–435. [CrossRef]
15. Sahoo, B.M.; Amgoth, T.; Pandey, H.M. Enhancing the Network Performance of Wireless Sensor Networks on Meta-heuristic Approach: Grey Wolf Optimization. In *Applications of Artificial Intelligence and Machine Learning. Lecture Notes in Electrical Engineering*; Choudhary, A., Agrawal, A.P., Logeswaran, R., Unhelkar, B., Eds.; Springer: Singapore, 2021; Volume 778, pp. 469–482. [CrossRef]
16. Verma, K.; Baliyan, N. Grey wolf optimization with fuzzy logic for energy-efficient communication in wireless sensor network-based Internet of Things scenario. *Int. J. Commun. Syst.* **2021**, *34*, e4981. [CrossRef]
17. Kapoor, R.; Sharma, S. Glowworm Swarm Optimization (GSO) based energy efficient clustered target coverage routing in Wireless Sensor Networks (WSNs). *Int. J. Syst. Assur. Eng. Manag.* **2021**, 1–13. [CrossRef]
18. Jabinian, Z.; Ayatollahitafti, V.; Safdarkhani, H. Energy Optimization in Wireless Sensor Networks Using Grey Wolf Optimizer. *J. Soft Comput. Decis. Support Syst.* **2018**, *5*, 1–6.
19. Jaiswal, K.; Anand, V. A QoS aware optimal node deployment in wireless sensor network using Grey wolf optimization approach for IoT applications. *Telecommun. Syst.* **2021**, *78*, 559–576. [CrossRef]
20. Saber, A.; Fekher, K.; Abbas, B.; Abderrezak, R.; Kaddachi, M.L.; Atri, M. A new fuzzy logic based node localization mechanism for Wireless Sensor Networks. *Future Gener. Comput. Syst.* **2019**, *93*, 799–813. [CrossRef]
21. Kaushik, A.; Indu, S.; Gupta, D. A Grey Wolf Optimization Approach for Improving the Performance of Wireless Sensor Networks. *Wirel. Pers. Commun.* **2019**, *106*, 1429–1449. [CrossRef]
22. Tay, M.; Senturk, A. A New Energy-Aware Cluster Head Selection Algorithm for Wireless Sensor Networks. *Wirel. Pers. Commun.* **2022**, *122*, 2235–2251. [CrossRef]
23. Jaya Pratha, S.; Asanambigai, V.; Mugunthan, S.R. Grey Wolf Optimization Based Energy Efficiency Management System for Wireless Sensor Networks. *Res. Sq.* **2021**. [CrossRef]
24. Lee, J.-G.; Chim, S.; Park, H.-H. Energy-Efficient Cluster-Head Selection for Wireless Sensor Networks Using Sampling-Based Spider Monkey Optimization. *Sensors* **2019**, *19*, 5281. [CrossRef]
25. Duraimurugan, S.; Avudaiammal, R. Energy Efficient Nodes Clustering and Routing Using Multi-Objective Spider Monkey Optimization Algorithm in Wireless Sensor Network. *Res. Sq.* **2021**. [CrossRef]
26. Rathore, R.S.; Sangwan, S.; Prakash, S. Hybrid WGWO: Whale grey wolf optimization-based novel energy-efficient clustering for EH-WSNs. *EURASIP J. Wirel. Commun. Netw.* **2020**, *101*, 101. [CrossRef]
27. Yue, Y.; You, H.; Wang, S.; Cao, L. Improved whale optimization algorithm and its application in heterogeneous wireless sensor networks. *Int. J. Distrib. Sens. Netw.* **2021**, *17*, 1–13. [CrossRef]
28. Agrawal, D.; Qureshi, M.H.W.; Pincha, P.; Srivastava, P.; Agarwal, S.; Tiwari; Pandey, S.V. GWO-C: Grey wolf optimizer-based clustering scheme for WSNs. *Int. J. Commun. Syst.* **2020**, *33*, e4344. [CrossRef]

29. Ghorpade, S.N.; Zennaro, M.; Chaudhari, B.S. Binary grey wolf optimisation-based topology control for WSNs. *IET Wirel. Sens. Syst.* **2019**, *9*, 333–339. [CrossRef]
30. Preeti, K.R.; Singh, D. Dimension learning based chimp optimizer for energy efficient wireless sensor networks. *Sci. Rep.* **2022**, *12*, 14968. [CrossRef]
31. Zachariah, U.E.; Kuppusamy, L. A hybrid approach to energy efficient clustering and routing in wireless sensor networks. *Evol. Intell.* **2021**, *15*, 593–605. [CrossRef]
32. Sackey, S.H.; Ansere, J.A.; Anajemba, J.H.; Kamal, M.; Iwendi, C. Energy Efficient Clustering Based Routing Technique in WSN using Brain Storm Optimization. In Proceedings of the 2019 15th International Conference on Emerging Technologies (ICET), Peshawar, Pakistan, 2–3 December 2019; pp. 1–6. [CrossRef]
33. Ali, H.; Tariq, U.U.; Hussain, M.; Lu, L.; Panneerselvam, J.; Zhai, X. ARSH-FATI a Novel Metaheuristic for Cluster Head Selection in Wireless Sensor Networks. *IEEE Syst. J.* **2021**, *15*, 2386–2397. [CrossRef]
34. Mirjalili, S.; Mirjalili, S.M.A. Lewis, Grey Wolf Optimizer. *Adv. Eng. Softw.* **2014**, *69*, 46–61. [CrossRef]
35. Albina, K.; Lee, S.G. Hybrid Stochastic Exploration Using Grey Wolf Optimizer and Coordinated Multi-Robot Exploration Algorithms. *IEEE Access* **2019**, *7*, 14246–14255. [CrossRef]
36. Alok, K.; Lekhraj; Safalata, S.; Anoj, K. Grey wolf optimizer and other metaheuristic optimization techniques with image processing as their applications: A review. *IOP Conf. Ser. Mater. Sci. Eng.* **2021**, *1136*, 012053. [CrossRef]
37. Anitha, P.; Kaarthick, B. RETRACTED ARTICLE: Oppositional based Laplacian grey wolf optimization algorithm with SVM for data mining in intrusion detection system. *J. Ambient. Intel. Humaniz. Comput.* **2021**, *12*, 3589–3600. [CrossRef]
38. Chaimaa, K.; Yahlal, M.; Boudıa, M.A.; Amıne, A.; Hamou, R.M.; Siham, K. Intrusion Detection System with Grey Wolf Optimizer (GWO). *Int. J. Inf. Appl. Math.* **2019**, *2*, 45–60.
39. Alzaqebah, A.; Aljarah, I.; Al-Kadi, O.; Damaševičius, R. A Modified Grey Wolf Optimization Algorithm for an Intrusion Detection System. *Mathematics* **2022**, *10*, 999. [CrossRef]
40. Yuvaraj, N.; Karthikeyan, T.; Praghash, K. An Improved Task Allocation Scheme in Serverless Computing Using Gray Wolf Optimization (GWO) Based Reinforcement Learning (RIL) Approach. *Wirel. Pers. Commun.* **2021**, *117*, 2403–2421. [CrossRef]
41. Jeniffer, J.T.; Chandrasekar, A. Optimal hybrid heat transfer search and grey wolf optimization-based homomorphic encryption model to assure security in cloud-based IoT environment. *Peer to Peer Netw. Appl.* **2022**, *15*, 703–723. [CrossRef]
42. Purushothaman, R.; Rajagopalan, S.P.; Dhandapani, G. Hybridizing Gray Wolf Optimization (GWO) with Grasshopper Optimization Algorithm (GOA) for text feature selection and clustering. *Appl. Soft Comput.* **2020**, *96*, 106651. [CrossRef]
43. Sharawi, M.; Emary, E. Impact of grey wolf optimization on WSN cluster formation and lifetime expansion. In Proceedings of the 2017 Ninth International Conference on Advanced Computational Intelligence, Doha, Qatar, 4–6 February 2017; pp. 157–162. [CrossRef]
44. Daneshvar, S.M.M.H.; Mohajer, P.A.A.; Mazinani, S.M. Energy-Efficient Routing in WSN: A Centralized Cluster-Based Approach via Grey Wolf Optimizer. *IEEE Access* **2019**, *7*, 170019–170031. [CrossRef]
45. Sekaran, K.; Rajakumar, R.; Dinesh, K.; Rajkumar, Y.; Latchoumi, T.P.; Kadry, S.; Lim, S. An energy-efficient cluster head selection in wireless sensor network using grey wolf optimization algorithm. *TELKOMNIKA Telecommun. Comput. Electron. Control* **2020**, *18*, 2822–2833. [CrossRef]
46. Wang, Z.; Xie, H.; Hu, Z.; Li, D.; Wang, J.; Liang, W. Node coverage optimization algorithm for wireless sensor networks based on improved grey wolf optimizer. *J. Algorithms Comput. Technol.* **2019**, *13*, 1–15. [CrossRef]
47. Amruta, L.; Damodar, R.E.; Venkatanareshbabu, K. Energy efficient load balancing approach for avoiding energy hole problem in WSN using Grey Wolf Optimizer with novel fitness function. *Appl. Soft Comput.* **2019**, *84*, 105706. [CrossRef]
48. Subramanian, P.; Sahayaraj, J.M.; Senthilkumar, S.; Alex, D.S. A Hybrid Grey Wolf and Crow Search Optimization Algorithm-Based Optimal Cluster Head Selection Scheme for Wireless Sensor Networks. *Wirel. Pers. Commun.* **2020**, *113*, 905–925. [CrossRef]
49. Al-Aboody, N.A.; Al-Raweshidy, H.S. Grey wolf optimization-based energy-efficient routing protocol for heterogeneous wireless sensor networks. In Proceedings of the 2016 4th International Symposium on Computational and Business Intelligence, Olten, Switzerland, 5–7 September 2016. [CrossRef]
50. Zhang, Y.; Cao, L.; Yue, Y.; Cai, Y.; Hang, B. A Novel Coverage Optimization Strategy Based on Grey Wolf Algorithm Optimized by Simulated Annealing for Wireless Sensor Networks. *Comput. Intel. Neurosci.* **2021**, *2021*, 1–14. [CrossRef]
51. Shipeng, W.; Xiaoping, Y.; Xingqiao, Y.; Zhihong, Q. A Virtual Force Algorithm-Lévy-Embedded Grey Wolf Optimization Algorithm for Wireless Sensor Network Coverage Optimization. *Sensors* **2019**, *19*, 2735. [CrossRef]
52. Saleh, I.A.; Alsaif, O.I.; Yahya, M.A. Optimal distributed decision in wireless sensor network using gray wolf optimization. *IAES Int. J. Artif. Intell.* **2020**, *9*, 646–654. [CrossRef]
53. Rajakumar, R.; Amudhavel, J.; Dhavachelvan, P.; Vengattaraman, T. GWO-LPWSN: Grey Wolf Optimization Algorithm for Node Localization Problem in Wireless Sensor Networks. *J. Comput. Netw. Commun.* **2017**, *2017*, 7348141. [CrossRef]
54. Liu, Y.; Jing, X.; Li, C.; Qin, H.; Jie, Z. Sensor Duty Cycle for Prolonging Network Lifetime Using Quantum Clone Grey Wolf Optimization Algorithm in Industrial Wireless Sensor Networks. *J. Sens.* **2021**, *2021*, 5511745. [CrossRef]
55. Hou, Y.; Gao, H.; Wang, Z.; Du, C. Improved Grey Wolf Optimization Algorithm and Application. *Sensors* **2022**, *22*, 3810. [CrossRef] [PubMed]
56. Li, Y.; Lin, X.; Liu, J. An Improved Gray Wolf Optimization Algorithm to Solve Engineering Problems. *Sustainability* **2021**, *13*, 3208. [CrossRef]

57. Kumar, A.; Kumar, A. Weighted Grey Wolf Optimizer with Improved Convergence rate in Training multi-layer Perception to solve Classification Problems. *Jordanian J. Computers Inf. Technol.* **2021**, *7*, 292–312. [CrossRef]
58. Nadimi-Shahraki, M.H.; Shokooh, T.; Seyedali, M. An improved grey wolf optimizer for solving engineering problems. *Expert Syst. Appl.* **2021**, *166*, 113917. [CrossRef]
59. Safaldin, M.; Otair, M.; Abualigah, L. Improved binary gray wolf optimizer and SVM for intrusion detection system in wireless sensor networks. *J. Ambient. Int. Humaniz. Comput.* **2021**, *12*, 1559–1576. [CrossRef]
60. Wang, J.S.; Li, S.X. An Improved Grey Wolf Optimizer Based on Differential Evolution and Elimination Mechanism. *Sci. Rep.* **2019**, *9*, 7181. [CrossRef]
61. Qiao, Y.; Hsu, H.Y.; Pan, J.S. Behavior-based grey wolf optimizer for a wireless sensor network deployment problem. *Intl. J. Ad Hoc Ubiquitous Comput.* **2022**, *39*, 70–82. [CrossRef]
62. Mittal, N.; Singh, U.; Salgotra, R.; Sohi, B.S. An energy efficient stable clustering approach using fuzzy extended grey wolf optimization algorithm for WSNs. *Wirel. Netw.* **2019**, *25*, 5151–5172. [CrossRef]
63. Yousif, Z.; Hussain, I.; Djahel, S.; Hadjadj-Aoul, Y. A Novel Energy-Efficient Clustering Algorithm for More Sustainable Wireless Sensor Networks Enabled Smart Cities Applications. *J. Sens. Actuator Netw.* **2021**, *10*, 50. [CrossRef]
64. Zhao, X.; Ren, S.; Quan, H.; Gao, Q. Routing Protocol for Heterogeneous Wireless Sensor Networks Based on a Modified Grey Wolf Optimizer. *Sensors* **2020**, *20*, 820. [CrossRef]
65. Zhao, X.; Zhu, H.; Aleksic, S.; Gao, Q. Energy-Efficient Routing Protocol for Wireless Sensor Networks Based on Improved Grey Wolf Optimizer. *KSII Trans. Internet Inf. Syst.* **2018**, *12*, 2644–2657. [CrossRef]

computers MDPI

Review

Blockchain-Based Internet of Things: Review, Current Trends, Applications, and Future Challenges

Tanweer Alam

Faculty of Computer and Information Systems, Islamic University of Madinah, Madinah 42351, Saudi Arabia; tanweer03@iu.edu.sa

Abstract: Advances in technology always had an impact on our lives. Several emerging technologies, most notably the Internet of Things (IoT) and blockchain, present transformative opportunities. The blockchain is a decentralized, transparent ledger for storing transaction data. By effectively establishing trust between nodes, it has the remarkable potential to design unique architectures for most enterprise applications. When it first appeared as a platform for anonymous cryptocurrency trading, such as Bitcoin, on a public network platform, blockchain piqued the interest of researchers. The chain is completed when each block connects to the previous block. The Internet of Things (IoT) is a network of networked devices that can exchange data and be managed and controlled via unique identifiers. Automation, wireless sensor networks, embedded systems, and control systems are just a few of the well-known technologies that power the IoT. Converging advancements in real-time analytics, machine learning, commodity sensors, and embedded systems demonstrate the rapid expansion of the IoT paradigm. The Internet of Things refers to the global networking of millions of networked smart gadgets that gather and exchange data. Integrating the IoT and blockchain technology would be a significant step toward developing a reliable, secure, and comprehensive method of storing data collected by smart devices. Internet-enabled devices in the IoT can send data to private blockchain networks, creating immutable records of all transaction history. As a result, these networks produce unchangeable logs of all transactions. This research looks at how blockchain technology and the Internet of Things interact to understand better how devices can communicate with one another. The blockchain-enabled Internet of Things architecture proposed in this article is a useful framework for integrating blockchain technology and the Internet of Things using the most cutting-edge tools and methods currently available. This article discusses the principles of blockchain-based IoT, consensus methods, reviews, difficulties, prospects, applications, trends, and communication between IoT nodes in an integrated framework.

Keywords: blockchain; smart contract; Internet of Things; security and privacy; proof of work

Citation: Alam, T. Blockchain-Based Internet of Things: Review, Current Trends, Applications, and Future Challenges. *Computers* **2023**, *12*, 6. https://doi.org/10.3390/computers12010006

Academic Editor: Sergio Correia

Received: 1 November 2022
Revised: 13 December 2022
Accepted: 20 December 2022
Published: 26 December 2022

1. Introduction

Stuart Haber and W Scott Stornetta discovered blockchain in the early 1990s [1]. Blockchain, the major Bitcoin cryptocurrency, was initially recognized in 2009. The blockchain evolved into a cryptocurrency platform. Commercial banks are the primary institutions they regard as a new price platform. However, since 2009, the blockchain aroused tremendous interest in various businesses. Consider the power of cutting in chain management, consumer procurement restrictions, and other areas to produce income with blockchain. Blockchain is a broad term for an open, unstable, public realm that allows anyone to securely transmit data using public key authentication and proof of work (PoW) because of its power-sharing traits, resilience, confidentiality, and book research abilities. Throughout the testing, the blockchain's usage was focused on its usability and speed difficulties. The authors assumed that five scientific arches were heavily used for the experiment: IEEEXplore, Springer, ACM digital library, and Scopus. After the first round was accomplished, only 150 articles were selected. After the first round, unusual papers were deleted, and 100 papers were selected. In the third round of paper selection, only 81 papers were selected. Blockchains, by definition, rely on a public directory. The public directory acts as a common transaction information database for many sites. The ledge was released in phases, and blocks were noted in the ledger. Blockchain is made up of blocks. Blockchain time is an indestructible database that is placed in a new time transaction and divided into a block hash chain. It details many copies of such blocks that were made and saved in the extracted device's protocol. The primary purpose is to locate a donor population known as miners. Blockchain technology Minors are linked to the current blockchain generation [2]. The technique argues that a few miners can amend a risky or inaccurate transaction at any time stamp. Various blockchain frameworks are now accessible, including public, non-public, licensing, and no blockchain authorization. Thanks to the social blockchain standard, anyone can subscribe to the series. Public blockchains are often authoritative, especially when all users have equal access rights and proportions. A personal blockchain ensures anonymity. On a personal blockchain, each player's tool is predetermined. Customers do not currently have equal access and equitable rights in the community as a result of these licenses. Bitcoin is a significant and still commonly used blockchain protocol. There may be a time constraint of up to 5 min, similar to how a new miner is normally unbiased and wants the best one to contribute or install a new blockchain device. The primary question in this circumstance is who will add the transactions and how. The same gain can be reached using two methods: send proof and process verification. Consider a case in which you want to pay. The foundation explains why signing transactions validates the use of cryptographic keys. In this scenario, community service providers or miners validate the legitimacy of digital signatures and ensure asset access. When these tasks are complete, the blockchain system delivers new enhancements [3]. Each block has its hash code, which includes the hash of the previous blocks in the series. It is also used to connect blocks in a specified manner. To build network leadership, everyone involved in mining must work hard and rapidly. Internal computing solves the problem of highlighting contradicting size computations for permanent length recordings. A leader can be elected in any situation. This leader enables several miners to attempt to rectify the problems with the person who solved the problem first and offer evidence of completion to the group. Furthermore, multiple miners must guarantee that the completed designs are ready. In contrast, confirming and picking that hole. Every node uses a blockchain verification mechanism before placing something in a blockchain device. If the nodes are authenticated blocks, they receive a pow. Every new node that joins a shared blockchain device receives a copy of the blockchain. When creating a new block, send it to all blockchain nodes remotely. Each node also validates the blocking process and ensures that the data displayed are correct. If the block is validated, it is posted to the nearby blockchain. Pos has another alternative. This is another approach. This approach selects the most useful person in the network. The Pos defines the value of the pole in the network dependent on the amount of money the miner has. This presupposes that a miner with several poles is very popular in the network [4]. All nodes eagerly greet this leader by including his block in the mining block. Only the inclusion of new in-network devices should keep the blockchain current. The research question is "what are the prospects and obstacles for a blockchain-based IoT network?".

There are various survey papers accessible on the blockchain. This study investigates opportunities, demands, and trends resulting from blockchain systems' complete functioning and behaviour. A few exam papers focus on specific features of the blockchain, such as framework, smart contracts, privacy, security, compliance agreements, applications, and IoT–blockchain integration. Few studies investigate the privacy and security challenges raised by specific blockchain topologies, such as Ethereum or Bitcoin. Table 1 displays similar studies from the same study area. The author specifies the role of blockchain in IoT, challenges in communication between blockchain and IoT, and the applications. The authors examine the characteristics of the current and future blockchain generation, significant performance difficulties, and open challenges for the future.

Table 1. Related Studies.

Reference(s)	Topics	Year	Area
[5]	Peer-to-Peer Electronic Cash System	2008	Bitcoin
[6]	Blockchain as a Service for IoT	2016	Blockchain and IoT
[7]	Blockchain Technology	2016	Blockchain
[8]	Blockchain for the Internet of Things	2016	Blockchain and IoT
[9]	Blockchain for the Internet of Things	2017	Blockchain and IoT
[10]	Internet of things and Blockchain	2016	Blockchain and IoT
[11]	Security of Blockchain	2017	Blockchain Security
[12]	Analysis of Established Blockchain Systems	2017	Blockchain
[13]	Consensus algorithms of Blockchain	2017	Blockchain
[14]	blockchain research framework	2019	Blockchain
[15]	Consensus protocols on Blockchain	2017	Blockchain
[16]	Blockchain framework	2018	Blockchain
[17]	Data processing view of Blockchain systems	2018	Blockchain
[18]	Blockchain: trends and future	2018	Blockchain
[19]	Blockchain security architecture for the Internet of Things	2018	Blockchain and IoT
[20]	Blockchain and IoT integration	2018	Blockchain and IoT
[21]	Use of Blockchain for the Internet of Things	2018	Blockchain and IoT
[22]	Blockchain meets IoT	2018	Blockchain and IoT
[23]	Blockchain technologies for the Internet of things	2018	Blockchain and IoT
[24]	The blockchain-empowered software system	2018	Blockchain
[25]	Distributed ledger technologies	2018	Blockchain
[26]	Blockchain Transactions	2018	Blockchain
[27]	Blockchain: Challenges and applications	2018	Blockchain
[28]	Blockchain applications in different domains	2020	Blockchain
[29]	Blockchain in developing countries	2018	Blockchain
[30]	Blockchain and its Role in the Internet of Things	2019	Blockchain and IoT
[31]	Blockchain applications in supply chain	2019	Blockchain
[32]	Privacy protection in blockchain system	2019	Blockchain
[33]	Blockchain in industries	2019	Blockchain
[34]	Blockchain for Internet of things	2019	Blockchain and IoT
[35]	Evolution of Blockchain	2019	Blockchain
[36]	Blockchain-based IoT	2019	Blockchain and IoT
[37]	Blockchain Technology	2019	Blockchain
[38]	Security Issues in IoT for Blockchain Healthcare	2019	Blockchain and IoT
[39]	Security issues and blockchain solutions for IoT	2020	Blockchain and IoT
[40]	Scalability of Blockchain Systems	2019	Blockchain
[41]	Security and privacy on the Blockchain	2019	Blockchain
[42]	Blockchain Applications for Industry 4.0	2019	Blockchain
[43]	Transformative effects of IoT, Blockchain and Artificial Intelligence	2019	Blockchain and IoT

Table 1. *Cont.*

Reference(s)	Topics	Year	Area
[44]	Efficiency Issues and Solutions in Blockchain	2019	Blockchain
[45]	Blockchain-based security aspects in heterogeneous Internet-of-Things	2018	Blockchain and IoT
[46]	Consensus Algorithms in Blockchain Technology	2019	Blockchain
[47]	Determining blockchain applicability	2018	Blockchain
[48]	Blockchain Applications in IoT Systems	2019	Blockchain and IoT
[49]	Aspects of Blockchain and IoT	2019	Blockchain and IoT
[50]	Blockchain characteristics and consensus	2019	Blockchain
[51]	Survey of Blockchain-Enabled Cyber-Physical Systems	2020	Cyber-Physical Systems
[52]	IoT Applications in Blockchain Systems	2020	Blockchain and IoT
[53]	Blockchain, Fog and IoT Integrated Framework	2020	Blockchain and IoT
[54]	Blockchain consensus algorithms performance evaluation	2020	Blockchain
[55]	Blockchain for 5G-enabled IoT	2020	Blockchain and IoT
[56]	Blockchain smart contracts formalization	2020	Blockchain
[57]	Blockchain for Cybersecurity in IoT	2021	Blockchain and IoT
[58]	Integration of Blockchain and Internet of Things: challenges and solutions.	2021	Blockchain and IoT
[59]	A Comprehensive Survey on Blockchain in Industrial Internet of Things: Motivations, Research Progresses, and Future Challenges	2022	Blockchain and IoT

The authors also described blockchain policy terms and changes to blockchain-related organizations. A scientific study is intended to reveal a target audience's range of behaviours and attitudes about important issues and concerns. The outcomes of qualitative research are explanatory rather than predictive. All information gathered consists of words written and uttered by people and their observed behaviours. In-depth interviews are a qualitative research approach used by researchers to collect data to acquire a more profound knowledge of the interviewee's perspective and situation. Such an interview strategy involves asking participants open-ended and screening questions to receive information that the researcher finds worthwhile. This article carefully explores how BC can benefit from the Internet of Things. The primary goal of this research is to analyze recent trends in BC-related approaches and tool usage analysis in an IoT environment. Compared to previous studies, this paper explores the novel roles of BC in IoT, finds new opportunities in various areas, for example, in the COVID-19 situation, and explores the challenges. Additionally, it shows the reader how far along numerous proposed solutions are in their advancement. We also discuss the main open questions, future research directions in the field, and the challenges the research community must overcome to integrate BC and IoT successfully.

This paper describes how blockchain technology can be applied to Internet of Things contexts to solve problems that arise. The following are some of the key contributions made by this study:

1. The paper begins with a brief introduction to the Internet of Things and blockchain. On the other side, this study reveals the numerous challenges experienced by researchers while exploring the Internet of Things.
2. This study highlights the importance of smart contracts in the Internet of Things environment.

3. Method of data storage and management, big data, cloud computing, and network security management technique are the three primary groups into which we categorize and investigate the available solutions in depth.
4. In the form of a table, we compare the categories of the offered solutions in terms of used technology, potential solutions, and implementation notes.
5. This paper covers unanswered research topics and our findings that may be applicable to the development of blockchain-based IoT systems, based on a review.

The following is how the rest of the article is organized. The authors present the broad structure and growing technology components of blockchain, including a distributed ledger, cryptography, consensus protocols, smart contracts, and standards, in Section 2. Section 3 depicts the consensus algorithms, Section 4 depicts the blockchain layered architecture, Section 5 defines current blockchain development trends, Section 6 depicts the role of blockchain in the IoT, Section 7 depicts communication among IoT nodes in the IoT–blockchain framework, Section 8 depicts challenges, and Section 9 depicts blockchain applications.

2. Blockchain

The blockchain is a public digital ledger constructed on a peer-to-peer network that may be openly distributed across varied users to generate an immutable history of timestamped and connected transactions [60]. When a group of transactions is appended, the data creates a new block in the chain. Simply put, a blockchain is a timestamped set of unchangeable data records attempted by a group of machines that a single party does not control. Each data block is encrypted and linked to the next in a chain using cryptography standards. Without relying on a centralized system, untrustworthy entities with common interests can utilize blockchain to construct a reliable, unchangeable, and open history of trading and distribution [61,62]. The numerous elements of blockchain and its reliance on encryption and distributed systems may allow for a more complex understanding.

Furthermore, each part may be clearly defined and used as a building brick to comprehend the greater complex structure. Blockchain technology can be characterized as public or private [63]. Anyone can connect to the blockchain network using the public blockchain. By confirming transactions and delivering correct services, all users contribute to the public blockchain. Public blockchains are the most extensively utilized cryptocurrencies today. Only authorized users have access to the private blockchain. The owner of the private blockchain can modify or remove entities from the blockchain network. A variety of corporate paradigms recommended the use of private blockchains. The blockchain is a distributed electronic database of digitally signed transactions that are clustered into chain blocks [13]. Each block is cryptographically connected to the preceding records after verification and consensus. When a new block is installed, reconfiguring earlier blocks is much more challenging. The new blocks are duplicated across all system copies of the ledger, and any conflicts are addressed instantly using current rules. The blockchain methods utilized to transmit data between devices are depicted in Figure 1.

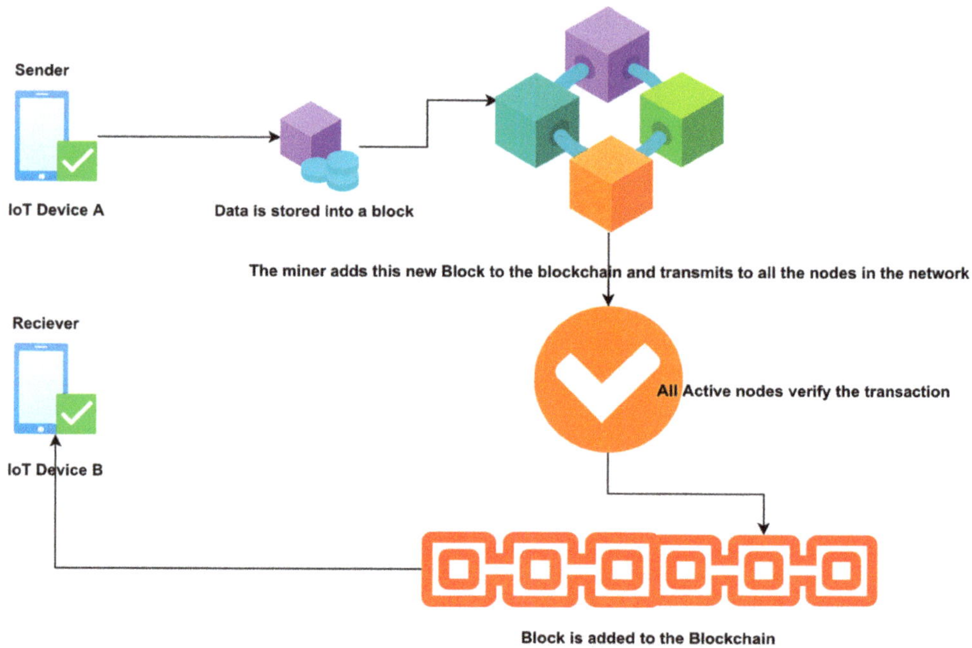

Figure 1. Blockchain process to send data between devices.

2.1. Components of Blockchain

The following are blockchain add-ons.

2.1.1. Block

A block is an information structure used to deceive a transaction group sent to all public nodes.

2.1.2. Nodes

Nodes for devices or users in the blockchain community.

2.1.3. Transactions

Transactions are the tiniest components of the blockchain framework design.

2.1.4. Miners

A miner is a particular node that executes a block authentication method on a blockchain network.

2.1.5. Chain

A chain is a sequence of blocks.

2.1.6. Consistency

Consistency refers to highly complex and fast rules that can be used to carry out blockchain operations when processing transactions. The blockchain can be represented as a collection of linked blocks. Before Bitcoin, it was known as digital ledgers in general. The blockchain is an actively distributed system that prioritizes data integrity, transparency, security, shareability and other features. To distribute processes across various locations, blockchain uses a distributed platform, often a peer-to-peer (P2P) system [64]. A consensus approach could be utilized to synchronize the saved and data gathered from each node [65].

Two prominent communication protocols are Gossip and Kademlia. Messages are sent over many endpoints to support these protocols. Gossip is the most widely used protocol in Bitcoin, and it is used to send data to the whole network, with each node only talking with its neighbours. Kademlia, on the other hand, defines the network structure using a shared hash table, and the peer list includes each node interaction.

2.2. Blockchain Versions

Table 2 depicts the evolution and versioning of blockchain technology from 1.0 to 4.0. With its first use (Bitcoin), blockchain 1.0 was introduced for distributed ledger technologies. Blockchain 2.0 introduced smart contracts, short computer programs that run automatically with verifications. Decentralized storage communication on peer-to-peer networks is possible with blockchain 3.0. Blockchain 4.0 enabled enterprises to access blockchain technology, allowing them to be employed in Industry 4.0. The current blockchain versions are listed in Table 2.

Table 2. Blockchain versions.

Year	Version	Application	Algorithms	Chaining	Execution Framework	Other Features
2008	1.0	Currency	PoW	Metachain	Bitcoin	Transparency, authentication, zminimize cost.
2013	2.0	Smart Contracts	PoW, PoS	Metachain	Ethereum	Distributed computations, Exchange the digital currencies
2015	3.0	Decentralized Apps	PoW, PoS, PoET, PBFT, etc.	A directed graph, Metachain, and sidechain.	Ethereum Swarm	Decentralized storage and communication
2018	4.0	Industry 4.0 Apps	Artificial intelligence-based Consensus	Connected chain, Divided chain	unibright.io framework	Approved workflows, financial transactions, IoT data gathering, e-health management system, etc.

2.3. Blockchain Terminologies

Blockchain terminologies vary from execution to execution—generic phrases would be utilized to communicate about the present invention. Some terminologies are defined below.

2.3.1. Blockchain

It is a distributed digital ledger.

2.3.2. Blockchain Technology

This term describes the innovation in the most generic version.

2.3.3. Blockchain Network

This term describes the network where a blockchain will be applied. It demonstrates blockchain implementation.

2.3.4. Blockchain Network User

This term describes an organization, individual, company, administration, and many others using the blockchain network system.

2.3.5. Node

A node is an individual device in the blockchain system.

2.4. Blocks in a Blockchain

Blockchain network users send transactions to the blockchain system through computer apps, smart device apps, digital wallets, web services, and other means. The apps route such transactions to a node or nodes within the blockchain system. These entire nodes could be published or unpublished. A submitted transaction would then be disseminated to other nodes in the network. However, they cannot find the transaction in the blockchain community. Because of the nature of many blockchain systems, non-time transactions must wait in a queue until a publishing node sends them to the blockchain network. Figure 2 illustrates the blockchain's blocks.

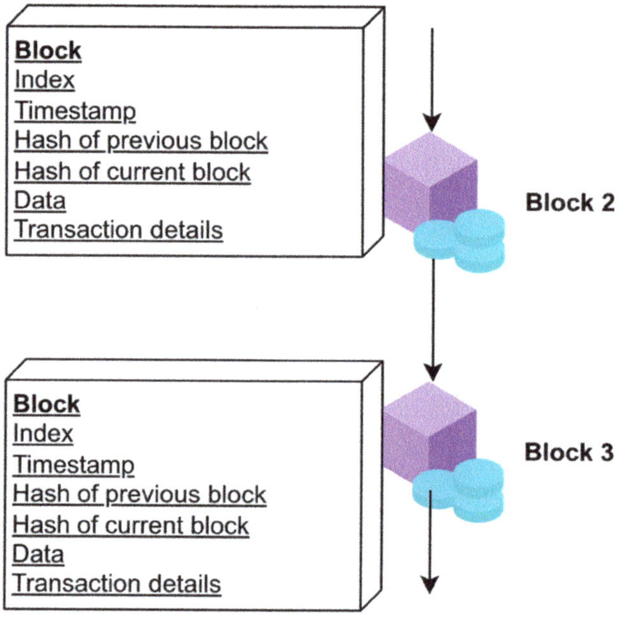

Figure 2. Blocks in blockchain.

A blockchain block might include both a header and information. This header consists of the metadata for the block. The data consists of a list of correct and confirmed transactions published in the blockchain system. Accuracy and validity are ensured by ensuring that the transaction is accurately configured and that the distributed ledger services cryptographically signed each transaction. In the blockchain implementation, there are no standard data fields for a block. They can describe their data fields in a block. However, data fields are used in a variety of blockchain implementations.

2.4.1. Block Header

In some blockchain networks, a block number is also referred to as a block height.

2.4.2. Hash Code

The hash code of the previous block pass. A hash description of the block information (this might be accomplished using a variety of approaches, such as creating a Merkle tree, processing the root hash, or utilizing a hash of all the combination block information).

2.4.3. Timestamp

A method for authenticating data and associating electronic files or events with a particular instant in time is known as timestamping, which is based on blockchain technology. A timestamp is a string of letters that serves as a unique identifier for the document or event in consideration and the time it was formed. In the most basic form, the timestamp is a string of letters.

2.4.4. Block Size

A nonce is something that is utilized for mining and is controlled by the publishing node to solve the hash puzzle.

Figure 3 depicts the process of connecting one block to two others, a previous and the following one, which validates the chain architecture of a blockchain. However, the question is, where does that chain begin, and where does it end? The solution is that the initial block, known as the genesis block, is hardcoded into the source code. This has a hash reference with only zeros and is the first block in the chain. It also contains certain arbitrary information that can be identified within its coin base transaction. The following is the block information:

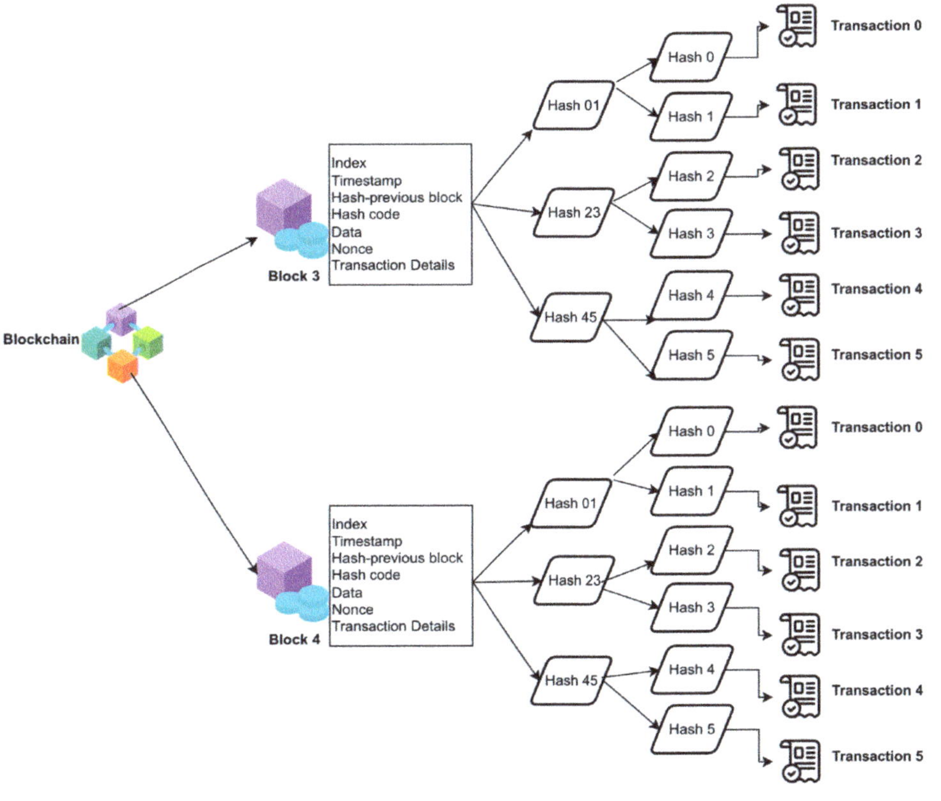

Figure 3. Blockchain structure.

2.4.5. Data

The information in the block.

2.4.6. The Ledger Transactions and Events

This section may include further information.

2.5. Transactions

A transaction is an activity that occurs between nodes. It depicts the exchange of digital currency between blockchain network nodes. In business-to-business contexts, it could be

3.1. Proof of Work (PoW)

PoW is the most common mechanism used for cryptocurrencies such as Bitcoin and Ethereum. Before proceeding, for non-technical users, the hashing method would be a way that may be used to assign random-sized material to fixed-size content. Whether a hash function is secure or not, its performance is random.

3.2. Proof of Stake (PoS)

Because the PoS does not allow devices to perform repeated processing, it is more environmentally sustainable. It replaces miners with verifiers, who can keep a portion of the Bitcoins as a stake. The team of verifiers takes turns suggesting and participating in the next block, and the stakes' quality determines the strength of each validator's voting. When the validators discover a block that they believe can be added to the blockchain, they can verify it by placing a bet on it as well. That validator would be given credit for their stakes. Everyone who maintains a cryptocurrency blockchain foundation can become a validator by submitting a specific type of operation to safeguard it. The following are some of the benefits of PoS.

(1) Altitude: allows for faster transactions.
(2) Efficiency: uses less power.
(3) Less equipment: no need for a supercomputer.

Its downside is vulnerability: anyone with enough money to invest exclusively in the destruction of this device might accomplish it by spending only money, as opposed to PoW, where they would have to spend money, time, skills, and so on.

3.3. Delegated Proof of Stake (DPoS)

Because delegated proof of stake (DPoS) and conventional PoS differ slightly, the gap between direct democracy and representative democracy can be compared. Any account containing Bitcoins is eligible to stake in a standard PoS system. People can vote on verifying activities, reach a shared consensus, and receive coins in exchange. Each wallet containing tokens is eligible to vote for representatives throughout the DPoS scheme [69].

3.4. Leased Proof of Stake (LPoS)

This method is an upgraded version of the stake consensus algorithm, which seeks to generate a decentralized consensus to secure the blockchain platform. This approach enabled regular people with no technical skills to contribute to the security of the Waves system by leasing Waves to complete peers without losing ownership of the keys. Meanwhile, the Waves blockchain enabled performance with access to 100 TPS, which is a completely different scenario than other blockchain technologies. Expenses are modest, and there is no need to provide miners with node incentives to compensate for high energy expenses and expensive equipment.

3.5. Proof of Elapsed Time (PoET)

Every participant will be given a network-authorized timing entity that will wait for the time provided by the timestamp. Instead, each participant obtains and sends a certification proving that the system stopped the remainder of the system. This network tests how frequently the participant is a member in order to determine the members. Every member of the blockchain system will delay for a set amount of time. The first member to complete the time restriction will become a participant in the newly created block. Each new member must gain access to the blockchain joining system; once activated, the policy will produce a new key pair, which the member will present to the entire network as part of a request to participate.

3.6. Practical Byzantine Fault Tolerance (PBFT)

This algorithm vastly improves aspects of Byzantine fault tolerance (in several terms, security towards Byzantine faults). It was already implemented in some significant decentralized computer networks and certain blockchain networks. The PoET stochastic selects a specific peer group to perform demands at a given moment. Typical observers must test a uniformly distributing random function and wait for the experiment's specified period. Its shortest test peers outperform it. Deception is avoided by using a secure implementation system, identification authentication, blocklisting centred on asymmetric key encryption, and a comprehensive set of laws.

3.7. Delegated Byzantine Fault Tolerance (DBFT)

This algorithm is utilized to get a consensus, which disappoints some blockchain and cryptocurrency developers. Delegated Byzantine fault tolerance is more efficient than most other methods in dealing with unstable or insecure blockchain participants.

3.8. Direct Acyclic Graph (DAG)

This is a topologically organized directed graph data model. Its series could only occur before and after. It is also used for data analysis, organization, choosing the best routing technique, and information decoding.

3.9. Proof of Activity (PoA)

This is referred to as a blockchain consensus algorithm that verifies that transactions are valid and that miners achieve an agreement. The PoA mixes PoW and PoS and attempts to bring the most powerful of either. Throughout POA, the refining process begins as a standard PoW mechanism, with multiple miners competing with a great computational capacity to surpass one another to find a new element. When a new (extracted) block is recognized, the device switches to POS, with the newly identified block consisting of the head and data.

3.10. Proof of Importance (PoI)

This became one of the most significant breakthroughs in blockchain-based businesses. This innovative approach employs system theory to assign a value to the channel for each consideration. Several additional blockchains assign incentives using proof of work (POW) or proof of stake (POS). Anyone who can maintain the most reliable connection arrays has an advantage over many other customers when it comes to POW. Such POW buildings frequently generate excessive amounts of power, harm the environment, and burden mining firms with high electricity bills. This gives card hoarders an unfair advantage. The more coins they use in transactions, the more money they receive.

3.11. Proof of Capacity (PoC)

Blockchain miners employ software backups rather than the more common power-efficient proof of work (PoW) technique, which combines continuous computer activity. The transaction system is protected throughout the proof of work method by performing an absurd amount of processing per moment to verify every block. That is why you must employ hardware. It usually results in negatives, such as increased electricity usage, temperature, loud noises, the requirement for modern semi-reusable equipment, and huge enterprises' centralization of the refining process.

3.12. Proof of Burn (PoB)

This algorithm is more attainable than others. It reduces energy consumption. There is no need to mine the hardware equipment in this algorithm. The coin burns are essentially virtual mining equipment. The PoB algorithm allows miner nodes to make long-term commitments with other nodes in the network. In this algorithm, the supply or extraction of coins looks to grow less centralized.

3.13. Proof of Weight (PoWeight)

This blockchain consensus mechanism assigns them a 'weight' based on the amount of digital currency they own. Consumers of PoWeight play an important role in the consensus mechanism. Each consumer is assigned a "weight" based on the relevance chosen to serve the customer's commitment to the system. It safeguards against double spending and other wrongdoing on the blockchain. The high proportion of the weighted portion must pose a risk to fair customers. The number of coins determines the likelihood of PoS at risk linked to the rest of the system. PoWeight, on the other hand, might employ different weighted standards to evaluate the possibilities.

4. Blockchain–IoT Layered Architecture

The blockchain comprises numerous components that perform various functions, such as processing transactions, propagating blocks, mining, finding approval, and storing the ledger for its related coins. The blockchain contains several levels similar to the well-known TCP/IP infrastructure. These elements could be classified based on their characteristics. There are still various concepts for creating a blockchain network with a tiered design. The blockchain layered architecture is depicted in Figure 6.

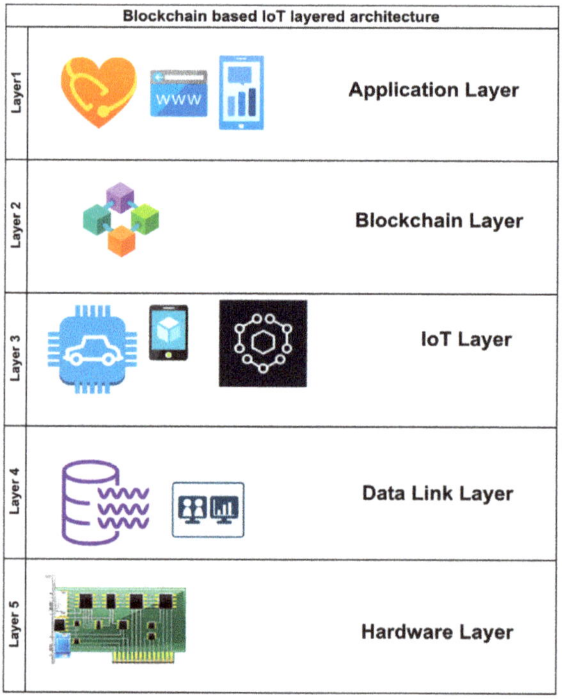

Figure 6. Blockchain layered architecture.

4.1. Application Layer

This layer includes smart contracts, chain code, and blockchain applications. This layer is divided into two sections: application and execution. In the blockchain system, end users use the applications to communicate and share information. Software, Web applications, user interfaces, and protocols are all included. For such applications, the blockchain system serves as the back-end infrastructure. These apps, however, frequently interact with the blockchain system via interfaces. The second tier is the implementation level, which includes smart contracts, basic rules, and a hybrid ledger. It follows strict coding and execution standards [70,71]. The smart contract was created in Solidity and will be executed on Ethereum's runtime platform. To show that coding requires the use of a compiler. The bytecode becomes shorter as it is compiled. As a result, it outperforms on Ethereum. The Ethereum software is isolated from the network and file system.

4.2. Blockchain Layer

Consensus is the most fundamental and necessary level of any blockchain. A consensus mechanism creates a proven collection of entity commitments in a decentralized P2P system. According to the consensus, most of the nodes are precisely aligned. The consensus mechanisms vary depending on the type of blockchain. The consensus mechanism is defined as deterministic if preceded by an unregulated blockchain network, such as Ethereum, Bitcoin, and so on. Although there is a danger that different parties may have different viewpoints on a block in the blockchains, consensus ensures the accuracy of the ledger. Deterministic approaches are used by permitted blockchains, such as Hyperledger. Unique nodes, known as ordered endpoints, exist on specific blockchain platforms.

4.3. Network Layer

This layer is accountable for transaction discovery and block distribution in the IoT environment. This means that nodes will discover each other and will be able to connect, share, and transfer data to better the blockchain system's existing position. The P2P platform is a network in which devices are shared and system loads are redistributed. Endpoints execute blockchain transactions. Full nodes and light nodes are the two types of nodes. Full nodes provide exchange verification, authentication, processing, and compliance with consensus laws. These nodes are in charge of ensuring the system's trust. The light node can convey the blockchain's header and submit operations.

4.4. Data Link Layer

The data structure of the blockchain could be regarded as a link list of blocks where transactions are structured. Merkle's tree is a hash code binary tree. All blocks contain a hash of Merkle's root and information, such as the hashing of the previous block, timestamp, nonce, block versions, and the most recent complexity level. Merkle trees, cryptography, and consensus mechanisms are the foundation for distributed ledger technology. The root hashing might be applied to the entire tree network. Every block lists multiple transactions that occurred since the previous transaction, and when those transactions are submitted, the root hashing indicates the current state of the blockchain.

4.5. Hardware Layer

Blockchains could be used to structurally measure, verify, and store transactions in a distributed database. This database contains all important information, transactions, and facts. A P2P system considers every computer to be an endpoint. Endpoints confirm transactions, organize them into blocks, upload them to the blockchain system, and so on. Nodes add blocks to the blockchain system and upgrade the public ledger replica after reaching an agreement. A virtualized layer is used to create such a layer. Substantially, the nodes are at the centre of this layer. A computer is referred to as a node when it is connected to a blockchain system. These nodes are part of a decentralized and distributed blockchain system.

5. Roles of Blockchain in IoT

Privacy and security in IoT device communication gained much attention between 2018 and 2022. Many papers will be published between 2018 and 2022 [72–76]. In 1990, Stuart The authors published an essay on sharing a record with privacy without storing any information on timestamping services. The blockchain concept originated with, although Satoshi Nakamoto provided the first blockchains in 2008. The author gave a document in which the blocks were joined to construct a blockchain in a certain order. The IoT chain is proposed by for data authentication between two nodes in IoT networks. In addition, they developed a set of guidelines for altering records in IoT and blockchain technology. The authors' goal is to secure the authorization component of the IoT chain device. The item researched the cloud and manet infrastructure to connect intelligent gadgets in the IoT and provide security and authentication during verbal transmission. It represents a suitable framework known as the net cloud framework, which could be useful for facilitating communication with IoT-based intelligent devices. It proposes a cloud-manet middleware system for accessing various IoT node records. The project also suggests ways to combine blockchains and the IoT. They also provide security within the blockchain–IoT to construct IoT apps using blockchain capability. The IoT enables connected physiological things in a heterogeneous system to modify their records. The Internet of Things would be divided into the following components.

5.1. Physical Things

The Internet of Things provides a unique identity for each related thing within the network device. IoT devices can exchange data with one another.

5.2. Gateways

Gateways are devices that function between physical objects and the cloud to guarantee the connection is secure and the network device is protected.

5.3. Networking

This is used to control the glide of statistics and determine the shortest path for various IoT devices.

5.4. Cloud

The cloud is used to store and compute records.

5.5. Storage

Blockchain is a collection of demonstrated and encrypted blocks of transactions stored on a network device. The records of blocks are recorded in a virtual ledger that is publicly shared, distributed, and open. Blockchain enables secure communication within the IoT community device. With extraordinary qualities, blockchain might be personal, public, or consortia.

5.6. Blockchain Ledgers

A decentralized trust model, excessive protection, extremely publicly accessible, privacy ranging from low to high, and transferable identities are all properties of blockchain ledgers [77,78]. The properties are centralized versions with low trust and public access. Privateness is an excessive and non-transferable acknowledgement. As a result, blockchain is more advanced than centralized storage systems.

5.7. Blockchains

Blockchains are a technological development that secures transactions between IoT devices. It enables decentralization, dissemination, and public peer-to-peer storage of sections of data stored or validated in an IoT network. The information recorded in the blockchain is quickly handled in a peer-to-peer setting. Blockchains are a technology that allows IoT devices to alter transactions in the blockchain device. The following summarizes blockchain's function in IoT.

5.8. Smart-Route Control Algorithm (s-RCA)

The s-RCA finds the optimal route for data traffic between a source and a destination, minimizing the total number of hops and the total amount of time it takes to transfer that data. This reliable connection can be used in medical procedures to guarantee that all instructions are received promptly and carried out without delay. This idea can enhance m-QoS for surgical procedures performed at a distance with the help of trusted paths. The new s-RCA can be integrated into an existing routing protocol to keep tabs on the primary path and track emergency packets stored in node buffers for immediate forwarding via the demand path [79].

5.9. Decentralized Framework

Both IoT and blockchain will be utilized in this framework. A structure such as this eliminates the need for a centralized approach and even allows for a decentralized design. It increases the likelihood of the cumulative control unit failing or performing poorly. The use of blockchain firmly established decentralization. A data offloading algorithm is also being developed to distribute various processing and computing tasks to OpenFlow switches based on their current load. A traffic model is also recommended for modelling and analyzing traffic in different network nodes. The proposed algorithm is evaluated using both a testbed and a simulation. The experimental results show that the proposed framework performs better regarding latency and resource consumption.

5.10. Exchanges between Nodes in Blockchains

Exchanges between nodes in blockchains are always secure. It is a novel way to achieve relationship security. Blockchains allow IoT devices to connect with one another more efficiently.

5.11. Identification

In the IoT, all connected devices are uniquely recognized using a cryptographic hash. Each block is also clearly labelled. As a result, blockchain is a powerful innovation that gives known facts that can be processed across the database.

5.12. Consistency

The blockchain database of nodes effectively acquired information scattered across the database. The data were correct when the miners double-checked it before entering it into the blockchain. Only validated blocks may enter the blockchain.

5.13. Autonomous

Every IoT device can communicate with any other device in a network via a distributed architecture.

5.14. Optimistic

IoT devices could interact with high availability, a decentralized intelligent network communicating with the target device in real-time, or transaction data.

6. Communication among IoT Nodes in an IoT Blockchain Framework

IoT blockchains evolved into a new ledger incorporating IoT users' utilized, used, public, and real-time performance. The blockchain is organized in blocks, and each block is linked to the blocks that came before it to form a chain. Every block contains a cryptographically secure token, a front hash block, and metadata. Blockchain transactions are basic gadgets used to send data between IoT-based devices. Nodes are various kinds of physical objects.

On the other hand, smart gadgets have embedded sensors, actuators, and packaging, and can interface with other IoT devices. Its role in the IoT is to offer secure system records to IoT devices. It is a luxury that may be publicly utilized and accepted by the public. The IoT requires this time to facilitate secure communication between IoT nodes in a heterogeneous machine. Anyone with authenticated communication within the IoT platform can track and monitor blockchain transactions. The IoT blockchain can increase chat security and authentication. Throughout this work, the author investigated this technique, its possibilities and situations that require assistance, and its thoughts.

6.1. Peer-to-Peer Network

Peers are computer systems linked to one another via the internet network in the peer-to-peer (P2P) ecosystem. Without a central system, documents could be shared instantly among devices on the platform. Any device connected to the P2P platform will function as a storage server and computer. The primary goal of P2P systems would be to communicate information and assist linked devices in interacting efficiently and effectively, gaining access to critical services, or performing specified duties. P2P is a network resource exchange protocol used to trade network resources, such as compute power, networking capacity, and data storage capacity. However, the most common use case for P2P networking is file sharing via a network. Peer-to-peer networks are ideal for file sharing because they allow connected devices to receive and transmit files simultaneously. Sensor nodes with P2P networking could play an essential role in the IoT system. A lightweight encryption technique is required for IoT sensor nodes.

Blockchains offer an efficient solution to a P2P communication network problem. This technology enables the creation of a shared electronic transaction ledger distributed among device peers rather than centralized. Users are linked to blockchains to track activity. This system employs cryptography to verify and recognize the participants' peers, allowing them to attach actions to the database cryptographically. Transactions are vetted and checked by specific nodes in the process, eliminating the need for a central system. This approach might be easily applied to IoT platforms to overcome the scale issue, allowing millions of devices to use the same infrastructure without requiring professional help. Blockchains also address the issue of authority conflicts among manufacturers by establishing a standard in which everyone has equal rights and privileges.

6.2. IoT-Blockchain Integration

Each IoT tool acts as a blockchain node, capable of generating transactions, sharing them to form a block as a miner, and acting as a more straightforward transaction validation node to establish a blockchain subgroup and confirm exchanges to apply a simplified change validation approach. In spite of the limited resources available, the integration of blockchain protocols into IoT devices is being carried out. Keeping and validating a large volume of blockchain data is especially useful when limited resources require sufficient recent memories or processing. As a result, the idea is that not all of the data collected by IoT devices should be kept and stored in the blockchain network. Almost all blockchain offerings, including IoT distribution networks, are open to the public. However, if the personal blockchain is used as much as feasible, scalability will no longer be as much of an issue as it is now [80]. Because the information is of the quickest subject, the private blockchain will no longer include that many performers. Its transmission is critical for the export company, the importer, and maybe other parties in the manufacturing process. Except for performers associated with the blockchain, most cannot participate in and verify the emergence.

6.3. IoT Blockchain Communication

The Internet of Things is rapidly expanding, with applications such as smart homes and cities, e-fitness, distributed intelligence, and so on. However, it creates difficult privacy and security conditions. To connect IoT devices, decentralized networking is used. As a result, employing the same old security approaches in spoken communication among IoT devices became significantly more complicated. Blockchain is a technology that provides security in transactions between different IoT nodes. Blockchain is a distributed, decentralized, and publicly accessible shared ledger that keeps track of the blocks processed and demonstrated in an IoT network. The public ledger's information is handled automatically via a peer-to-peer network. Blockchain is a generation in which IoT device transactions are recorded as a block in the blockchain. Blocks are linked, and each tool has a tool reference that comes before it. The approaches to blockchain and IoT integration work within IoT and cloud integration techniques. Blockchain has the potential to alter future IoT conversations. The visions of blockchain and IoT integration are outlined below.

1. The decentralized method is quite similar to IoT and blockchain technologies. This removes the centralized device and provided the power of a decentralized method. This reduces the likelihood of failure and enhances the overall performance of the framework.
2. Security: Blockchain enables secure transactions between nodes. This is a revolutionary communication strategy. The Blockchain enables IoT devices to communicate with one another in a safe environment.
3. Identifications: IoT assists all associated gadgets that are uniquely recognized with a unique id variation. Every block in a blockchain is also uniquely identified.
4. However, blockchain is a trusted era that gives uniquely recognized information kept in a shared public ledger.
5. Reliability: The IoT nodes in blockchain can authenticate the information passed over the networks. Facts are reliable because miners validate them before entering the blockchain system. However, only the most useful proved blocks can be included in the blockchain device.
6. Autonomous: The blockchain enables all IoT nodes to connect with any node in the network without relying on a centralized approach.
7. Scalability: Blockchain enables IoT devices to communicate in a distributed intelligence network. It also communicates with real-time destination tools and alternate facts.

6.4. Platforms

Several systems are used to create IoT packages using blockchain.

1. IoTa: IoTa is the new platform for blockchain and IoT, also known as next-generation blockchain. By utilizing fewer assets within the device, the platform contributes to high information integrity, overall transaction performance, and block validity. This also resolves the blockchain restrictions.
2. IoTify: this provides a web-based IoT approach to reduce the constraints of blockchain in the form of customized applications.
3. Iexec: this open-source blockchain-based device is used to assist your apps and the blockchain's decentralized cloud benefits.
4. Xage: this versatile blockchain platform for IoT allows for increased automation and more relaxed data in the machine.
5. SONM is a decentralized blockchain-based fog computing platform that simplifies cloud offerings for users.

The IoT and blockchain are extending company potential and introducing new marketplaces in which anybody or everything can connect in real-time in a decentralized device with authenticity, privacy, and security. Incorporating these revolutionary technologies will revolutionize the current world on numerous levels, with gadgets communicating without people. The framework's goal is to provide safe data at the right place, in a suitable format, and real-time on a device. Blockchain will fine-tune billions of IoT-connected concerns, coordinate this stuff, facilitate transaction processing, resolve or eliminate crises, and build a flexible environment for running physical things. Blockchain develops data privacy for clients connected to the framework by utilizing hashing procedures in information blocks.

7. Current Trends in BC-IoT Development

Due to the security method of sending transactions between numerous entities without a trusted third and monitoring information veracity, blockchain gained a lot of attention. Even though many analysts feel that blockchain is the solution to many problems in today's fundamentally insecure internet because of its privacy and security capabilities, there may not be a systematic study to examine and evaluate blockchain from various angles [81]. Blockchain made its debut in 2009, less than a decade earlier. Because of this extraordinary innovation, the globe underwent a rapid transition. Blockchain is making inroads into various professional industries, including retail, medical care, and science.

7.1. Federated Blockchain

Federated blockchain is one of the most significant and successful recent blockchain innovations. This is an upgraded method of the basic blockchain framework, making it perfect for various relevant applications. According to experts, federated blockchain could grow in popularity since it provides a more customizable perspective for private blockchain. Federated blockchains are comparable to private blockchains in most ways, with some small advantages. These blockchains are speedier (higher scalability) and provide greater transaction privacy. Federated blockchain examples include R3 (banks), EWF (energy), B3i (insurance), Corda, and more.

7.2. Blockchain as a Service (BaaS)

BaaS is the creation and maintenance by third parties of cloud-based systems used by industries to run blockchain-based apps. Another trend that Microsoft and Amazon are utilizing is BaaS. This recent blockchain trend is currently integrated with various startups and businesses. However, such future blockchain trends may not be viable while building, sustaining, and monitoring a new blockchain technique. It is a cloud-based system that uses blockchain technology to enable users to develop online services. Such digital products could be smart contracts, apps, or other services that operate independently of the blockchain-based network.

7.3. Ricardian Contracts

The Ricardian contract was designed to identify a legally valid document that was electronically linked to an important aspect. The Ricardian contract organizes all of the legal agreement facts into a layout that the program can execute. As a result, it is both a legal contract and a mechanism that electronically integrates the agreement into digital infrastructure while providing a secure network because of cryptographic verification. It is distinct from the smart contract. Smart contracts are a type of digital agreement that was previously agreed upon and is automatically executable. On the other hand, the Ricardian contract is an agreement paradigm for recording an agreement's objectives and any behavior related to that agreement prior to the agreement being performed. Ricardian contracts could likewise be simply applied to software by utilizing hashes that describe external documentation.

7.4. Blockchain Interoperability

Interoperability refers to exchanging information and other content across many blockchain networks and infrastructures. The public could easily access information that was held on a number of different blockchains because of this function [82,83]. It enables subscribers to transfer funds easily and quickly from one blockchain to another. This functionality also adds additional functions, such as cross-chain transactions. It can also improve multi-token transactions by constructing multi-token wallet services.

7.5. Social Networking

Blockchain in social networks would be capable of resolving difficulties such as prominent debates, privacy violations, information manipulation, and the significance of the material. As a result, blockchain is a new technology trend that is being integrated into social media architecture. Tokens are used by social networks. As a result, media companies are given financial incentives to generate content and increase network productivity. Token exchanges, such as the blockchain, are finished and practically instantaneous, with no charges.

7.6. Hybrid Blockchains

The future scope of blockchain technology, which could be simplified as the blockchain, proposes using a more appropriate fraction of public and private blockchain technologies. The exchange rate is slightly lower because the network's popular nodes make validating processes simple and quick. The hybrid blockchain operates in a closed ecosystem, so all evidence on the network improves security. This also avoids more than half of all attacks, since thieves cannot gain access to the blockchain system. The user can update the rules whenever it is necessary. It also helps to keep a task secret when engaging with the outside world.

8. Opportunities within the Integrated Technique

Several excellent potentials for blockchain–IoT integration were discovered. Blockchain–IoT together opens new opportunities in several regions. Figure 7 depicts the blockchain–IoT potential. However, some of the options are listed here.

Figure 7. Blockchain–IoT opportunities.

8.1. Create Trust among Gadgets

Because of its security, blockchain-IoT technology will instill trust in some of the numerous connected devices. However, only the most basic verified devices can communicate within the community, and the miners must first validate each transaction block before it can enter the blockchain [84].

8.2. Reduce the Expenses

It will save money because it communicates instantly using an online platform. This removes all third-party nodes. It also allows direct communication among IoT nodes.

8.3. Reduce Time

It may drastically shorten the time. Blockchain-IoT cuts transaction time from days to seconds.

8.4. Security and Privacy

Blockchain-IoT provides devices with security and anonymity, and data are exchanged between devices [85].

8.5. Social Services

Blockchain-IoT provides public and social services to connected nodes. Each connected gadget can communicate and exchange data [86].

8.6. Financial Services

It can safely transfer payments without the involvement of other parties. Blockchain-IoT enables a quick, secure, and private financial services system. It also decreases transfer costs and time, among other things [87].

8.7. Risk Management

It played a critical role in analyzing and mitigating the risk of system resources and transactions failing [88].

9. Challenges

Many difficulties, including measurement, storage, services, and discovery, could be addressed by blockchain–IoT. The blockchain–IoT challenges are depicted in Figure 8. The integrated strategy faces the following challenges.

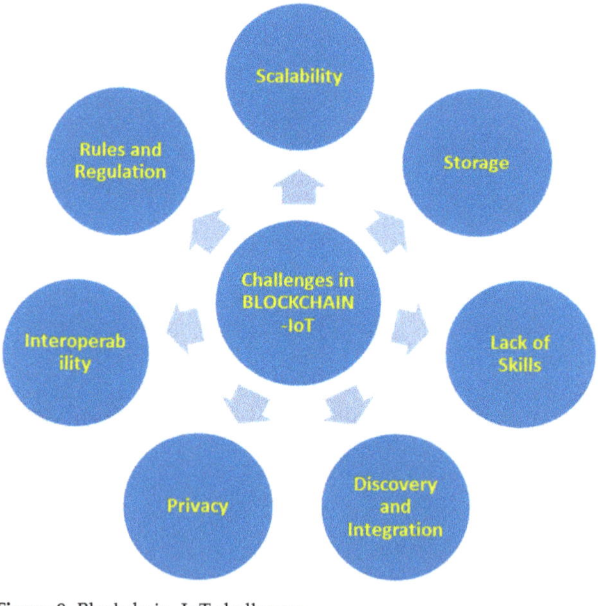

Figure 8. Blockchain–IoT challenges.

9.1. Scalability

Because of its high transaction volume, blockchain can become blocked. On 10 July 2022, Bitcoin had more than 406 GB of storage space [89]. Weight can be considerably more significant than the best blockchain when IoT joins blockchain.

9.2. Storage

All IoT devices will store the given ledger. However, it will increase with its storage size in the form of being challenging to work with and a significant weight on every connected tool.

9.3. Inadequate Abilities

Most researchers are unaware of the blockchain phenomenon, which is increasingly mainstream. It is also an initiative to teach people nearly every technology.

9.4. Exploration and Integration

Blockchain is not intended for IoT applications. It is quite tricky for associated devices to locate another tool within the blockchain and IoT systems. As a result, IoT nodes can access all other nodes while detecting and integrating blockchain and other nodes.

9.5. Confidentiality

The shared ledger is broadcast to all connected devices. These gadgets are capable of viewing ledger transactions in real-time. As a result, maintaining privacy in an embedded system is a difficult task [90].

9.6. Interaction

Blockchain might be public, private, or in the form of a consortium. As a result, the interaction between public and private blockchains is also a blockchain IoT agreement.

9.7. Rules and Regulations

Because the IoT blockchain will function internationally, numerous norms and laws will be enforced globally.

This research looks on a novel approach known as blockchain-IoT. The article [91] discusses numerous possibilities and challenges. Similarly, this article lists the platforms that are available to implement Blockchain-IoT technique. Blockchain-IoT has the potential to lead the internet by redesigning and replacing it with a new internet service in which each smart device connects to other peer-to-peer devices. This can save both money and time while providing precise data to the relevant device. As a result, it has the potential to be a very useful strategy in the future.

10. Applications

The modern internet is concerned with the availability and security of connected resources. These resources might be encrypted on a network-to-network chain known as the blockchain or ledger, where each user knows with whom they transact. Because it simplifies the business, speeds up the process, eliminates failures, and saves it, it may safeguard commercial connections and avoid fraud. The distributed blockchain technology will transform people's life by allowing them to execute trades or control money via phones, vote, rent a car, and even prove their identity.

10.1. Smart Devices

A smart device connects wirelessly and gives users more excellent knowledge and control than ever before. For instance, if your washing machine stops working, a code associated with your device can connect to the internet and warn you. Such alerts keep the devices in good working order, saving them money on energy efficiency and allowing you to monitor the gadgets while on the road to work. Accessing such gadgets via the blockchain would safeguard the assets while allowing for information flow.

10.2. Sensors for the Supply Chain

A sensor is a device that detects and responds to a specific sort of input derived from physical infrastructure. Light, wind, movement, humidity, strain, or any other environmental changes may provide critical information. Sensors in the supply chain aid in the location of vehicle temperatures, pressures, and so on. When supply chain executives need to watch product or vehicle situations and determine precisely where they are and where they are heading, such inputs are used. The value of sensors in the supply chain is based on their ability to provide real-time activity information.

10.3. The Smart Contract

The smart contract evolved into a digital machine designed to electronically encourage, test, or execute the arrangement and execution of an agreement. Smart contracts allow for the performance of trustworthy exchanges without the use of third-party providers.

10.4. Keeping Track of Prescription Medications

Blockchains could improve the patient experience by allowing them to scan a barcode and instantaneously determine whether a prescription is counterfeit, according to the release. Its innovation could also determine when pharmaceuticals were gathered and transferred across the production process at the required temperature changes.

10.5. Voting through Electronic Means

In any country, the security of a vote is a matter of national security. Computer security is investigating the prospect of using an online voting system to lower the cost of hosting a federal election while meeting and increasing security criteria. Since the inception of democratic politicians, the voting mechanism was based on paper and pen. It is critical to replace the current pen-and-paper procedure with modern election technology to reduce fraud and make voting verifiable and trackable. Blockchains enable a wide range of applications that profit from the exchange. Blockchain has the potential to play a significant role in implementing shared electronic voting systems as a service. However, establishing electronic voting systems and using blockchain to implement these structures is fraught with difficulties.

10.6. Healthcare on the Blockchain

The blockchain can revolutionize healthcare data, place the patient at the centre of health infrastructure, and improve healthcare security, privacy, transparency, and connectivity. Such innovation has the potential to create a new architecture for health information sharing by making electronic records more influential, without an intermediary, and secure. This modern, continuously changing environment is ideal for innovation, research, and proof of concept testing.

10.7. Blockchain Music

In some cases, blockchain technology may be advantageous in music. It could increase availability or incentivize and share revenue with viewers through special edition electronic releases. However, employing it for a music service is unjust, and claiming it is a solution to any of the most pressing issues musicians confront is false.

10.8. Blockchain Identification

Identity verification is now an important aspect of our daily life. Visiting another nation, purchasing a new automobile, and enrolling in a university necessitate identification checks. Creating a new social media account also necessitates mobile authorization. Bringing personal belongings may not always be practical or even possible. That is where the blockchain's effective confidentiality control comes into play.

10.9. Passports

The idea behind a blockchain-based passport is that citizens can control electronic transport identification using provable information, such as biometrics, travel history, or any other related information gathered at checkpoints by trusted public authorities and other contact points. Instead of keeping the details to themselves, each authority or agency decides to rely on each other's data. Blockchain passport enables partnership members to obtain verifiable traveller-identifying information claims to assess validity, streamline passenger processing, and reduce risks. People can also use blockchain passport to retain their identification and acquire personal information, and digital attestations choose which data to transmit. The more attestations a traveller possesses, the more alliance members, governments, and other parties may be able to guarantee smooth and secure travel.

10.10. Certificates of Birth, Marriage, and Death

Some things are more significant than showing the paperwork, such as birth certificates, marriage certificates, and expiry certificates that give you access to various benefits (such as elections, employment, and residency), but ineptitude is becoming more widespread. According to UNICEF, more than one-third of children under five do not have a birth certificate. The Blockchain can make records more secure by obtaining birth and death certificates and allowing users to access this vital information.

10.11. Processing of Insurance Claims

Smart contracts on the blockchain network might be used to handle insurance claims. In this case, all parties to an insurance policy may have access to the shared insurance ledger to view policy details. When a claim is applied, a claimant can submit evidence such as insurance papers, claim documents, and supporting claims proof to the distributed ledger. For statements, policyholders must interact directly with distributors. This activity is recorded on a private blockchain, with smart contracts enabling a workflow claim. Blockchain financial services and tariff plans need to be scrutinized. The active policy and smart contracts will be placed on the blockchain for policyholders with preset claim requirements.

10.12. Data Exchange

Blockchain is primarily concerned with improving the efficiency of data exchange across the supply chain, including producers, shipping providers, distributors, governments, suppliers, fulfilment centres, and consumers. Blockchains will allow the corporation to track the source of degradation much more quickly, reducing the impact of tainted products. Regarding customer refunds, blockchains can give end-to-end information traceability, the best right to examine the product's background, and real-time position and condition.

10.13. Copyright and Royalties Are Protected

Blockchains could be game changers for copyright holders looking to defend their rights electronically. Without question, it began to make its presence known to copyright holders. It remains to be seen whether the compliance procedures recommended for such networks will be implemented. However, the outlook remains positive. It is difficult to imagine blockchain being utilized to secure copyright in the coming months. This technique must first be broadly adopted before it can be widely used to defend copyright.

10.14. Property Registration, Real Estate, and Land Registration

Blockchains can profoundly alter the real estate market, from property acquisition to title management. It can transform the relationship between taxpayers and tax authorities and change how tax returns are submitted, taxes are paid, and data are handled. Blockchain technology can disrupt and restructure finance and streamline transaction, exchange, and property registration processes.

10.15. In a Catastrophic Situation (COVID-19)

The COVID-19 pandemic emphasizes global interconnectedness. This also highlights a complex reality: vast amounts of critical information stay trapped in fortified information storage facilities and reputation mechanisms when we require swift, collective action or cooperation. The blockchain–IoT integrated solutions aid in the resolution of the most difficult issues confronting us between 2019 and 22.

11. Conclusions

A critical component of this decentralization strategy was the blockchain design, which included hash-based proof of work, shared key encryption, and peer-to-peer networks. Complexity, limited compatibility, resource constraints, privacy and security concerns, and vulnerabilities hamper current IoT solutions. The rapid advancement of blockchain technology provides solutions to problems such as increased connectivity, privacy, security, transparency, and stability. Academics investigate the intersection of blockchain and the IoT throughout this post. They also discussed and presented literature on blockchain and the IoT. The issues and applications for developing a stable and interoperable communication infrastructure for blockchain and the IoT are discussed. This article examines current blockchain trends. The integration of blockchain and IoT architecture is investigated, as are the advantages and disadvantages of the combined strategy.

Funding: This research received no external funding.

Data Availability Statement: Not applicable.

Conflicts of Interest: The author declares no conflict of interest.

Abbreviations

IoT	Internet of Things
P2P	Peer-to-peer
PoS	Proof of stake
PoW	Proof of work
R3	An enterprise blockchain technology company
EWF	Energy web foundation
B3i	The blockchain insurance industry initiative
Corda	Open-source blockchain platform for business
Chain	A sequence of blocks
DPoS	Delegated proof of stake
PBFT	Practical Byzantine fault tolerance
dBFT	Delegated Byzantine fault tolerance
LPoS	Leased proof of stake
PoET	Proof of elapsed time
DBFT	Delegated Byzantine fault tolerance
DAG	Direct acyclic graph
POA	Proof of activity
PoI	Proof of importance
PoC	Proof of capacity
PoB	Proof of burn
PoWeight	Proof of weight
IOTA	The next generation of distributed ledger technology
IoTIFY	Online cloud-based MQTT/HTTP network simulator
iExec	Blockchain-based decentralized cloud computing
Xage	Blockchain cybersecurity system
SONM	Decentralized fog computing platform

References

1. Haber, S.; Stornetta, W.S. How to timestamp a digital document. In Proceedings of the Conference on the Theory and Application of Cryptography, Aarhus, Denmark, 21–24 May 1990; Springer: Berlin/Heidelberg, Germany, 1990; pp. 437–455. [CrossRef]
2. Uddin, M.A.; Stranieri, A.; Gondal, I.; Balasubramanian, V. A Survey on the Adoption of Blockchain in IoT: Challenges and Solutions. *Blockchain Res. Appl.* **2021**, *2*, 100006. [CrossRef]
3. Alangot, B.; Achuthan, K. Trace and track: Enhanced pharma supply chain infrastructure to prevent fraud. In Proceedings of the International Conference on Ubiquitous Communications and Network Computing, Bangalore, India, 3–5 August 2017; Springer: Cham, Switzerland, 2017; pp. 189–195. [CrossRef]
4. Andoni, M.; Robu, V.; Flynn, D.; Abram, S.; Geach, D.; Jenkins, D.; McCallum, P.; Peacock, A. Blockchain technology in the energy sector: A systematic review of challenges and opportunities. *Renew. Sustain. Energy Rev.* **2019**, *100*, 143–174. [CrossRef]

5. Nakamoto, S. Bitcoin: A Peer-to-Peer Electronic Cash System. 2008. Available online: https://bitcoin.org/bitcoin.pdf (accessed on 23 October 2022).
6. Samaniego, M.; Jamsrandorj, U.; Deters, R. Blockchain as a Service for IoT. In Proceedings of the 2016 IEEE International Conference on Internet of Things (iThings) and IEEE Green Computing and Communications (GreenCom) and IEEE Cyber, Physical and Social Computing (CPSCom) and IEEE Smart Data (SmartData), Chengdu, China, 15–18 December 2016; pp. 433–436. [CrossRef]
7. Ammous, S. *Blockchain Technology: What is it good for?* SSRN: Rochester, NY, USA, 2016. [CrossRef]
8. Conoscenti, M.; Vetro, A.; De Martin, J.C. Blockchain for the Internet of Things: A systematic literature review. In Proceedings of the 2016 IEEE/ACS 13th International Conference of Computer Systems and Applications (AICCSA), Agadir, Morocco, 29 November–2 December 2016; pp. 1–6. [CrossRef]
9. Risius, M.; Spohrer, K. A blockchain research framework. *Bus. Inf. Syst. Eng.* **2017**, *59*, 385–409. [CrossRef]
10. Huckle, S.; Bhattacharya, R.; White, M.; Beloff, N. Internet of things, Blockchain and shared economy applications. *Procedia Comput. Sci.* **2016**, *98*, 461–466. [CrossRef]
11. Xia, B.; Ji, D.; Yao, G. Enhanced tls handshake authentication with blockchain and smart contract (short paper). In *International Workshop on Security*; Springer: Cham, Switzerland, 2017; pp. 56–66. [CrossRef]
12. Haffke, F. Technical Analysis of Established Blockchain Systems. Master's Thesis, Technical University of Munich, SW Engineering for Business Informatics, München, Germany, 2017.
13. Mingxiao, D.; Xiaofeng, M.; Zhe, Z.; Xiangwei, W.; Qijun, C. A review on consensus algorithm of blockchain. In Proceedings of the 2017 IEEE International Conference on Systems, Man, and Cybernetics (SMC), Banff, AB, Canada, 5–8 October 2017; pp. 2567–2572. [CrossRef]
14. Wang, S.; Ouyang, L.; Yuan, Y.; Ni, X.; Han, X.; Wang, F.Y. Blockchain-enabled smart contracts: Architecture, applications, and future trends. *IEEE Trans. Syst. Man Cybern. Syst.* **2019**, *49*, 2266–2277. [CrossRef]
15. Sankar, L.S.; Sindhu, M.; Sethumadhavan, M. Survey of consensus protocols on blockchain applications. In Proceedings of the 2017 4th International Conference on Advanced Computing and Communication Systems (ICACCS), Coimbatore, India, 6–7 January 2017; pp. 1–5. [CrossRef]
16. Marsal-Llacuna, M.L. Future living framework: Is Blockchain the next enabling network? Technol. *Forecast. Soc. Change* **2018**, *128*, 226–234. [CrossRef]
17. Dinh, T.T.A.; Liu, R.; Zhang, M.; Chen, G.; Ooi, B.C.; Wang, J. Untangling Blockchain: A data processing view of blockchain systems. *IEEE Trans. Knowl. Data Eng.* **2018**, *30*, 1366–1385. [CrossRef]
18. Yang, W.; Garg, S.; Raza, A.; Herbert, D.; Kang, B. Blockchain: Trends and future. In *Pacific Rim Knowledge Acquisition Workshop*; Springer: Cham, Switzerland, 2018; pp. 201–210. [CrossRef]
19. Alphand, O.; Amoretti, M.; Claeys, T.; Dall'Asta, S.; Duda, A.; Ferrari, G.; Rousseau, F.; Tourancheau, B.; Veltri, L.; Zanichelli, F. IoTChain: A blockchain security architecture for the Internet of Things. In Proceedings of the 2018 IEEE Wireless Communications and Networking Conference (WCNC), Barcelona, Spain, 15–18 April 2018; pp. 1–6. [CrossRef]
20. Panarello, A.; Tapas, N.; Merlino, G.; Longo, F.; Puliafito, A. Blockchain and IoT integration: A systematic survey. *Sensors* **2018**, *18*, 2575. [CrossRef]
21. Fernández-Caramés, T.M.; Fraga-Lamas, P. A Review on the Use of Blockchain for the Internet of Things. *IEEE Access* **2018**, *6*, 32979–33001. [CrossRef]
22. Novo, O. Blockchain meets IoT: An architecture for scalable access management in IoT. *IEEE Internet Things J.* **2018**, *5*, 1184–1195. [CrossRef]
23. Ferrag, M.A.; Derdour, M.; Mukherjee, M.; Derhab, A.; Maglaras, L.; Janicke, H. Blockchain technologies for the Internet of things: Research issues and challenges. *IEEE Internet Things J.* **2018**, *6*, 2188–2204. [CrossRef]
24. Cai, W.; Wang, Z.; Ernst, J.B.; Hong, Z.; Feng, C.; Leung, V.C. Decentralized applications: The blockchain-empowered software system. *IEEE Access* **2018**, *6*, 53019–53033. [CrossRef]
25. El Ioini, N.; Pahl, C. A review of distributed ledger technologies. In Proceedings of the OTM Confederated International Conferences "On the Move to Meaningful Internet Systems", Valletta, Malta, 22–26 October 2018; Springer: Cham, Switzerland, 2018; pp. 277–288. [CrossRef]
26. Shrestha, A.K.; Vassileva, J. Bitcoin Blockchain Transactions Visualization. In Proceedings of the 2018 International Conference on Cloud Computing, Big Data and Blockchain (ICCBB), Fuzhou, China, 15–17 November 2018; pp. 1–6. [CrossRef]
27. Tasatanattakool, P.; Techapanupreeda, C. Blockchain: Challenges and applications. In Proceedings of the 2018 International Conference on Information Networking (ICOIN), Chiang Mai, Thailand, 10–12 January 2018; pp. 473–475. [CrossRef]
28. Chang, M.C.; Park, D. How can Blockchain help people in the event of pandemics such as the COVID-19? *J. Med. Syst.* **2020**, *44*, 102. [CrossRef] [PubMed]
29. Kshetri, N.; Voas, J. Blockchain in developing countries. *It Prof.* **2018**, *20*, 11–14. [CrossRef]
30. Alam, T. IoT-Fog: A Communication Framework using Blockchain in the Internet of Things. *Int. J. Recent Technol. Eng.* **2019**, *7*. [CrossRef]
31. Dujak, D.; Sajter, D. Blockchain applications in supply chain. In *SMART Supply Network*; Springer: Cham, Switzerland, 2019; pp. 21–46. [CrossRef]

32. Feng, Q.; He, D.; Zeadally, S.; Khan, M.K.; Kumar, N. A survey on privacy protection in blockchain system. *J. Netw. Comput. Appl.* **2019**, *126*, 45–58. [CrossRef]
33. Al-Jaroodi, J.; Mohamed, N. Blockchain in industries: A survey. *IEEE Access* **2019**, *7*, 36500–36515. [CrossRef]
34. Dai, H.N.; Zheng, Z.; Zhang, Y. Blockchain for Internet of Things: A survey. *IEEE Internet Things J.* **2019**, *6*, 8076–8094. [CrossRef]
35. Dabbagh, M.; Sookhak, M.; Safa, N.S. The evolution of Blockchain: A bibliometric study. *IEEE Access* **2019**, *7*, 19212–19221. [CrossRef]
36. Thakore, R.; Vaghashiya, R.; Patel, C.; Doshi, N. Blockchain-based IoT: A survey. *Procedia Comput. Sci.* **2019**, *155*, 704–709. [CrossRef]
37. Mohanta, B.K.; Jena, D.; Panda, S.S.; Sobhanayak, S. Blockchain technology: A survey on applications and security privacy challenges. *Internet Things* **2019**, *8*, 100107. [CrossRef]
38. Devibala, A. A Survey on Security Issues in IoT for Blockchain Healthcare. In Proceedings of the 2019 IEEE International Conference on Electrical, Computer and Communication Technologies (ICECCT), Coimbatore, India, 20–22 February 2019; pp. 1–7. [CrossRef]
39. Sengupta, J.; Ruj, S.; Bit, S.D. A comprehensive survey on attacks, security issues and blockchain solutions for IoT and IIoT. *J. Netw. Comput. Appl.* **2020**, *149*, 102481. [CrossRef]
40. Xie, J.; Yu, F.R.; Huang, T.; Xie, R.; Liu, J.; Liu, Y. A survey on the scalability of blockchain systems. *IEEE Network* **2019**, *33*, 166–173. [CrossRef]
41. Zhang, R.; Xue, R.; Liu, L. Security and privacy on blockchain. *ACM Comput. Surv. (CSUR)* **2019**, *52*, 1–34. [CrossRef]
42. Alladi, T.; Chamola, V.; Parizi, R.M.; Choo, K.K.R. Blockchain applications for industry 4.0 and industrial IoT: A review. *IEEE Access* **2019**, *7*, 176935–176951. [CrossRef]
43. Gill, S.S.; Tuli, S.; Xu, M.; Singh, I.; Singh, K.V.; Lindsay, D.; Tuli, S.; Smirnova, D.; Singh, M.; Jain, U.; et al. Transformative effects of IoT, Blockchain and Artificial Intelligence on cloud computing: Evolution, vision, trends and open challenges. *Internet Things* **2019**, *8*, 100118. [CrossRef]
44. Odiljon, A.; Gai, K. Efficiency Issues and Solutions in Blockchain: A Survey. In Proceedings of the International Conference on Smart Blockchain, Birmingham, UK, 11–13 October 2019; Springer: Cham, Switzerland, 2019; pp. 76–86. [CrossRef]
45. Pohrmen, F.H.; Das, R.K.; Khongbuh, W.; Saha, G. Blockchain-based security aspects in Internet of Things network. In Proceedings of the International Conference on Advanced Informatics for Computing Research, Shimla, India, 14–15 July 2018; Springer: Singapore, 2018; pp. 346–357. [CrossRef]
46. Sharma, K.; Jain, D. Consensus algorithms in blockchain technology: A survey. In Proceedings of the 2019 10th International Conference on Computing, Communication and Networking Technologies (ICCCNT), Kanpur, India, 6–8 July 2019; pp. 1–7. [CrossRef]
47. Scriber, B.A. A framework for determining blockchain applicability. *IEEE Softw.* **2018**, *35*, 70–77. [CrossRef]
48. Noby, D.A.; Khattab, A. A Survey of Blockchain Applications in IoT Systems. In Proceedings of the 2019 14th International Conference on Computer Engineering and Systems (ICCES), Cairo, Egypt, 17–18 December 2019; pp. 83–87. [CrossRef]
49. Atlam, H.F.; Wills, G.B. Technical aspects of Blockchain and IoT. In *Advances in Computers*; Elsevier: Amsterdam, The Netherlands, 2019; Volume 115, pp. 1–39. [CrossRef]
50. Viriyasitavat, W.; Hoonsopon, D. Blockchain characteristics and consensus in modern business processes. *J. Ind. Inf. Integr.* **2019**, *13*, 32–39. [CrossRef]
51. Rathore, H.; Mohamed, A.; Guizani, M. A survey of Blockchain enabled cyber-physical systems. *Sensors* **2020**, *20*, 282. [CrossRef]
52. Lao, L.; Li, Z.; Hou, S.; Xiao, B.; Guo, S.; Yang, Y. A survey of IoT applications in blockchain systems: Architecture, consensus, and traffic modeling. *ACM Comput. Surv.* **2020**, *53*, 18. [CrossRef]
53. Alam, T.; Benaida, M. CICS: Cloud-internet communication security framework for the internet of smart devices. *Int. J. Interact. Mob. Technol.* **2018**, 12. [CrossRef]
54. Bamakan, S.M.H.; Motavali, A.; Bondarti, A.B. A survey of blockchain consensus algorithms performance evaluation criteria. *Expert Syst. Appl.* **2020**, *154*, 113385. [CrossRef]
55. Mistry, I.; Tanwar, S.; Tyagi, S.; Kumar, N. Blockchain for 5G-enabled IoT for industrial automation: A systematic review, solutions, and challenges. *Mech. Syst. Signal Process.* **2020**, *135*, 106382. [CrossRef]
56. Singh, A.; Parizi, R.M.; Zhang, Q.; Choo KK, R.; Dehghantanha, A. Blockchain smart contracts formalization: Approaches and challenges to address vulnerabilities. *Comput. Secur.* **2020**, *88*, 101654. [CrossRef]
57. Chentouf, F.Z.; Bouchkaren, S. Blockchain for Cybersecurity in IoT. In *Artificial Intelligence and Blockchain for Future Cybersecurity Applications*; Springer: Cham, Switzerland, 2021; pp. 61–83. [CrossRef]
58. Zafar, S.; Bhatti, K.M.; Shabbir, M.; Hashmat, F.; Akbar, A.H. Integration of blockchain and Internet of Things: Challenges and solutions. *Ann. Telecommun.* **2022**, *77*, 13–32. [CrossRef]
59. Huo, R.; Zeng, S.; Wang, Z.; Shang, J.; Chen, W.; Huang, T.; Wang, S.; Yu, F.R.; Liu, Y. A Comprehensive Survey on Blockchain in Industrial Internet of Things: Motivations, Research Progresses, and Future Challenges. *IEEE Commun. Surv. Tutor.* **2022**, *24*, 88–122. [CrossRef]
60. Bansod, S.; Ragha, L. Blockchain Technology: Applications and Research Challenges. In Proceedings of the 2020 International Conference for Emerging Technology (INCET), Belgaum, India, 5–7 June 2020; pp. 1–6. [CrossRef]

61. Cervone, L.; Palmirani, M.; Vitali, F. The Intelligible Contract. In Proceedings of the HICSS, Maui, HI, USA, 7–10 January 2020; pp. 1–10. [CrossRef]
62. Cui, Z.; Fei XU, E.; Zhang, S.; Cai, X.; Cao, Y.; Zhang, W.; Chen, J. A hybrid BlockChain-based identity authentication scheme for multi-WSN. *IEEE Trans. Serv. Comput.* **2020**, *13*, 241–251. [CrossRef]
63. Saxena, S.; Bhushan, B.; Ahad, M.A. Blockchain based solutions to secure IoT: Background, integration trends and a way forward. *J. Netw. Comput. Appl.* **2021**, *181*, 103050. [CrossRef]
64. Dunphy, P.; Petitcolas, F.A. A first look at identity management schemes on the Blockchain. *IEEE Secur. Priv.* **2018**, *16*, 20–29. [CrossRef]
65. Ekramifard, A.; Amintoosi, H.; Seno, A.H.; Dehghantanha, A.; Parizi, R.M. A systematic literature review of integration of Blockchain and artificial intelligence. In *Blockchain Cybersecurity, Trust and Privacy*; Springer: Berlin/Heidelberg, Germany, 2020; pp. 147–160. [CrossRef]
66. Guidi, B. When Blockchain meets online social networks. *Pervasive Mob. Comput.* **2020**, *62*, 101131. [CrossRef]
67. Kshetri, N.; Voas, J. Blockchain-enabled e-voting. *IEEE Softw.* **2018**, *35*, 95–99. [CrossRef]
68. Lu, Y. The Blockchain: State-of-the-art and research challenges. *J. Ind. Inf. Integr.* **2019**, *15*, 80–90. [CrossRef]
69. Notheisen, B.; Cholewa, J.B.; Shanmugam, A.P. Trading real-world assets on Blockchain. *Bus. Inf. Syst. Eng.* **2017**, *59*, 425–440. [CrossRef]
70. Oham, C.; Jurdak, R.; Kanhere, S.S.; Dorri, A.; Jha, S. B-fica: Blockchain based framework for auto-insurance claim and adjudication. In Proceedings of the 2018 IEEE International Conference on Internet of Things (iThings) and IEEE Green Computing and Communications (GreenCom) and IEEE Cyber, Physical and Social Computing (CPSCom) and IEEE Smart Data (SmartData), Halifax, NS, Canada, 30 July–3 August 2018; pp. 1171–1180. [CrossRef]
71. Perrons, R.K.; Cosby, T. Applying Blockchain in the geoenergy domain: The road to interoperability and standards. *Appl. Energy* **2020**, *262*, 114545. [CrossRef]
72. Alam, T. Blockchain-based big data integrity service framework for IoT devices data processing in smart cities. *Mindanao J. Sci. Technol.* 2021. [CrossRef]
73. Alam, T. Blockchain-Enabled Deep Reinforcement Learning Approach for Performance Optimization on the Internet of Things. *Wirel. Pers. Commun.* **2022**, *126*, 995–1011. [CrossRef]
74. Alam, T. Cloud-based IoT applications and their roles in smart cities. *Smart Cities* **2021**, *4*, 1196–1219. [CrossRef]
75. Alam, T. IoT-fog-blockchain framework: Opportunities and challenges. In *Research Anthology on Convergence of Blockchain, Internet of Things, and Security*; IGI Global: Hershey, PA, USA, 2023; pp. 258–277. [CrossRef]
76. Alam, T.; Ullah, A.; Benaida, M. Deep reinforcement learning approach for computation offloading in blockchain-enabled communications systems. *J. Ambient Intell. Humaniz. Comput.* **2022**, 1–14. [CrossRef]
77. Tang, B.; Kang, H.; Fan, J.; Li, Q.; Sandhu, R. IoT passport: A blockchain-based trust framework for collaborative internet-of-things. In Proceedings of the 24th ACM Symposium on Access Control Models and Technologies, Toronto, ON, Canada, 3–6 June 2019; pp. 83–92. [CrossRef]
78. Uriarte, R.B.; DeNicola, R. Blockchain-based decentralized cloud/fog solutions: Challenges, opportunities, and standards. *IEEE Commun. Stand. Mag.* **2018**, *2*, 22–28. [CrossRef]
79. Abujassar, R.S.; Yaseen, H.; Al-Adwan, A.S. A Highly Effective Route for Real-Time Traffic Using an IoT Smart Algorithm for Tele-Surgery Using 5G Networks. *J. Sens. Actuator Netw.* **2021**, *10*, 30. [CrossRef]
80. Li, X.; Jiang, P.; Chen, T.; Luo, X.; Wen, Q. A survey on the security of blockchain systems. *Future Gener. Comput. Syst.* **2020**, *107*, 841–853. [CrossRef]
81. Salah, K.; Rehman MH, U.; Nizamuddin, N.; Al-Fuqaha, A. Blockchain for AI: Review and open research challenges. *IEEE Access* **2019**, *7*, 10127–10149. [CrossRef]
82. Savelyev, A. Copyright in the blockchain era: Promises and challenges. *Comput. Law Secur. Rev.* **2018**, *34*, 550–561. [CrossRef]
83. Sullivan, C.; Burger, E. E-residency and Blockchain. *Comput. Law Secur. Rev.* **2017**, *33*, 470–481. [CrossRef]
84. Liu, Y.; Yu, F.R.; Li, X.; Ji, H.; Leung, V.C. Blockchain and machine learning for communications and networking systems. *IEEE Commun. Surv. Tutor.* **2020**, *22*, 1392–1431. [CrossRef]
85. Vashisht, S.; Gaba, S.; Dahiya, S.; Kaushik, K. Security and Privacy Issues in IoT Systems Using Blockchain. In *Sustainable and Advanced Applications of Blockchain in Smart Computational Technologies*; Chapman and Hall/CRC: Boca Raton, FL, USA, 2022; pp. 113–127. [CrossRef]
86. Dahiya, A.; Gupta, B.B.; Alhalabi, W.; Ulrichd, K. A comprehensive analysis of Blockchain and its applications in intelligent systems based on IoT, cloud and social media. *Int. J. Intell. Syst.* 2022. [CrossRef]
87. Elngar, A.A.; Kayed, M.; Emira, H.H.A. The role of Blockchain in financial applications: Architecture, benefit, and challenges. In *Artificial Intelligence and Big Data for Financial Risk Management*; Routledge: London, UK, 2022; pp. 140–159. [CrossRef]
88. Choudhary, T.; Virmani, C.; Juneja, D. Convergence of Blockchain and IoT: An Edge Over Technologies. In *Toward Social Internet of Things (SIoT): Enabling Technologies, Architectures and Applications*; Springer: Cham, Switzerland, 2020; pp. 299–316. [CrossRef]
89. Statista. Size of the Bitcoin Blockchain from January 2009 to 11 July 2022. 2022. Available online: https://www.statista.com/statistics/647523/worldwide-bitcoin-blockchain-size/ (accessed on 21 October 2022).

90. Ebrahim, M.; Hafid, A.; Elie, E. Blockchain as privacy and security solution for smart environments: A Survey. *arXiv* **2022**, arXiv:2203.08901.
91. Conti, M.; Kumar, E.S.; Lal, C.; Ruj, S. A survey on security and privacy issues of bitcoin. *IEEE Commun. Surv. Tutor.* **2018**, *20*, 3416–3452. [CrossRef]

computers

MDPI

Communication

Energy Efficiency of IoT Networks for Environmental Parameters of Bulgarian Cities

Zlatin Zlatev [1,*], Tsvetelina Georgieva [2], Apostol Todorov [1] and Vanya Stoykova [1]

[1] Faculty of Technics and Technologies, Trakia University, 38 Graf Ignatiev Str., Yambol, 8602 Stara Zagora, Bulgaria; apostol.todorov.17@trakia-uni.bg (A.T.); vanya.stoykova@trakia-uni.bg (V.S.)
[2] Department of Automatics and Mechatronics, Ruse University, "Angel Kanchev" Ruse 8 Studentska Str., 7017 Ruse, Bulgaria; cgeorgieva@uni-ruse.bg
* Correspondence: zlatin.zlatev@trakia-uni.bg

Abstract: Building modern Internet of Things (IoT) systems is associated with a number of challenges. One of the most significant among them is the need for wireless technology, which will serve to build connectivity between the individual components of this technology. In the larger cities of Bulgaria, measures to ensure low levels of harmful emissions, reduce noise levels, and ensure comfort in urban environments have been taken. LoRa technology shows more advantages in transmission distance and low energy consumption compared to other technologies. That is why this technology was chosen for the design of wireless sensor networks (WSN) for six cities in Bulgaria. These networks have the potential to be used in IoT configurations. Appropriate modules and devices for building WSN for cities in Bulgaria have been selected. It has been found that the greater number of nodes in the WSN leads to an increase in the average power consumed in the network. On the other hand, depending on the location of these nodes, the energy consumed may decrease. The performance of wireless sensor networks can be optimized by applying appropriate routing protocols, which are proposed in the available literature. The methodology for energy efficiency analysis of WSN can be used in the design of wireless sensor networks to determine the parameters of the environment, with the possibility of application in IoT.

Keywords: wireless sensor networks; node localization; security; energy efficiency; IoT analysis; LoRa

Citation: Zlatev, Z.; Georgieva, T.; Todorov, A.; Stoykova, V. Energy Efficiency of IoT Networks for Environmental Parameters of Bulgarian Cities. *Computers* **2022**, *11*, 81. https://doi.org/10.3390/computers11050081

Academic Editor: Sergio Correia

Received: 24 April 2022
Accepted: 13 May 2022
Published: 17 May 2022

1. Introduction

The need for networking of non-complex sensor devices is increasing worldwide [1] in connection with the application of the Internet of Things, which is based on technologies designed for networking, low-cost, low-power consumption, and low data transfer capacity.

When creating a WSN to monitor the parameters of the urban environment, we are not looking for direct profit, but to improve people's living conditions. Using sensor data, it is possible to assess what needs to be changed and where, for example by regulating and controlling car traffic so that we can breathe cleaner air. Funding comes from the state (represented by municipal authorities). One of their duties is to take care of people's health. In the same way, funds from European environmental programs and projects can be used.

In automating the processes of measuring, transmitting, storing and processing data on the state of the environment in urban conditions, IoT technology is used in [2]: monitoring of environmental pollutants such as vehicles, industrial enterprises; climate change; disaster and accident warnings; assistance to people with disabilities; detection of aging and defects in machinery and equipment; advertising and media; evaluation of products and services; public safety in the city.

As a result of the measurements and analysis of the data received from the measuring stations, the governing bodies in the city organize activities to improve the health of the environment in cities. Such activities include a change in the organization of road transport in the city in order to reduce noise and the amount of harmful substances. In the same way, recommendations can be generated for production enterprises in the city to reduce harmful emissions and waste products resulting from the production they perform. The ultimate goal to be achieved is to create healthy living conditions for the population and protect the environment by developing and implementing an integrated approach and measures to avoid air and water pollutants and high noise levels by preventing or reducing these harmful effects.

In recent years, measures have been adopted in Bulgaria to measure the levels of harmful emissions, noise and pollutants from transport and industrial enterprises in urban environments. These measures are in line with the requirements of the European Union to ensure a healthy living environment in cities. The potential applications of IoT in this regard are diverse. The European Commission's IoT documents include the so-called smart city, defined as "a place where traditional networks and services become more efficient by using digital and telecommunications technologies for the benefit of residents and businesses" [3].

Updating the map for noise in settlements in Bulgaria is related to the fulfillment of the requirements of the Environmental Noise Protection Act (EPAA) (Promulgated, SG No. 74 of 13 September 2005) and Directive 2002/49/EC for the assessment and management of environmental noise. According to these requirements, it is the duty of every agglomeration with a population of over 100,000 inhabitants to update its noise map.

In the larger cities of the Republic of Bulgaria such measures to ensure low levels of harmful emissions, reduce noise levels, and ensure comfort in urban environments have been taken [4]. The measures are also supported by companies offering mobile services [5,6]. The following are provided [5]: intelligent parking systems, which show drivers the nearest free parking space and navigate them to it; intelligent lighting, providing greater security for citizens as well as saving on electricity costs for municipalities; air quality control; effective management of garbage containers; remote and intelligent metering of water meters, electricity meters and other measuring devices for households.

Measurements are based on automatic specialized stations, as well as by measuring with portable measuring instruments. The trend is to expand networks for automatic measurement and storage. For this purpose, it is necessary to select the appropriate technology for the wireless transmission of sensor data.

A review of the available literature sources [1,4,6] shows that in Bulgaria there is a need to create new and expand existing wireless sensor networks to measure the levels of pollutants in the environment. There is little research on the energy efficiency of this type of network.

Building modern systems based on IoT is associated with a number of challenges. One of the most important among them is the need for wireless technology, which will serve to build connectivity between the individual components of a new technology. There are various standards for wireless communication. Several of them are widely used and applicable to objects and devices with extremely low power. They can be IP-compatible or non-IP-compliant. Most of the wireless standards applicable to low-power devices are characterized by having a short range [7,8]. These standards are suitable for use in the construction of WSN, of course, taking into account their advantages and limitations.

The simulation analyses presented in some of the available literature [9,10] sources are related to "hypothetical" WSNs, which have a certain radius of availability and operating parameters. It is necessary to make studies which take into account the working conditions, geographical location, and number of nodes in the WSN.

Table 1 provides a comparative analysis of the performance of four commonly used wireless communication technologies that are suitable for use in sensor networks. The difference in bandwidth between the compared wireless communication technologies is not large. Baker [9] points out that the ZigBee over 802.15.4 protocol has more application capabilities than Bluetooth due to its long battery life, greater usable range, dimensional flexibility, and reliability of the network architecture. According to Lee et al. [10], the effective use of network protocols depends to a large extent on specific practical applications, as well as network reliability, roaming capability, recovery mechanism, hardware cost and network installation and maintenance costs. LoRa technology shows more advantages in transmission distance and a lower power consumption than other technologies. Due to the potentially complex environment in which the sensor devices will operate, problems also arise when using LoRa technology. In an urban environment, these are the presence of buildings, trees, and features of the terrain that prevent the transmission of signals in wireless networks [11].

Table 1. Standards for wireless data transmission.

Standard	Range	Topology	Frequency	Speed	Energy Consumption	Practical Implementation
IEEE 802.15.1 WPAN (Bluetooth)	100 m	Star, Mesh	2.4 GHz	1–24 Mb/s	Low	Sensor data transfer
IEEE 802.15.4 LRWPAN (ZigBee)	100 m	Star, Tree, Mesh	2.4 GHz	250 Kb/s	Low	Sensor data transfer
2G/3G	10 km	Star	Cellular network	10 Mb/s	High	Mobile, Data transfer
4G/5G	10 km	Star	5G–28 GHz; 4G-700- 2500 MHz	Up to 10 Gb/s	High	Mobile, IoT
Sigfox (0G)	10 km	Star	868–869 MHz, 902–928 MHz	Up to 0.6 Kbit/s	Medium	Sensor data transfer
IEEE 802.11 WLAN (WiFi)	1.5 km	Star	2.4, 3.6, 4.9, 5 GHz	11 Mb/s–1 Gb/s	Medium	Mobile, Data transfer
LoRa	15 km	Star	433, 868, 915 MHz	27 Kb/s	Very low	Sensor data transfer, IoT

The LoRaWAN protocol (https://www.thethingsnetwork.org, accessed on 1 April 2022) is built on LoRa technology. LoRa provides the physical layer that allows for long-distance bonding. LoRaWAN defines the communication protocol and system architecture. Protocol and network architecture have the greatest impact on battery life at a node, network capacity, quality of service, security, and the variety of applications served by the network. If the sensors need to be used as portable devices, this requires a communication protocol for dynamic LoRa sensor networks. This protocol is "multi-hop". An advantage of this transmission method is that restrictions on the maximum distances at which WSN data are received/transmitted can be avoided. The limitations of multi-hop are that it has high latency and low reliability.

When setting up communication in a LoRa network, the following key characteristics need to be taken into account [12]: Transmission Power (TP), which varies in the range [−4, 20 dBm], but due to some restrictions can be used in the range 2–20 dBm; Carrier Frequency (CF), which changes in steps of 61 Hz, in the range 137–1020 MHz, for which the scope depends on the scheme used and the region in which it is used; Spreading Factor (SF), which represents the ratio between the symbol rate and the chip rate and varies in the range of 6–12. High levels of this factor increase the signal-to-noise ratio (SNR), as well as increase the rate of transmission and time to transfer data packets; Bandwidth (BW), which represents the bandwidth of the transmission band. At higher BW values, more data are transmitted simultaneously, but at the expense of lower sensitivity and noise overlay in the signal. It is commonly used at frequencies of 125, 250 and 500 kHz; Code Rate (CR), which is a method of error correction. Higher CR levels allow for better noise reduction but reduce data rates. The default is 4/8, but it can have values of 4/5, 4/6, 4/7, or 4/8.

The aim of this paper is to clarify the technical challenges related to the energy efficiency of using LoRa networks in six cities in Bulgaria.

The main contributions that can be summarized are: a comparative analysis of WSN for six cities in Bulgaria; when designing a WSN, it is necessary to take into account the number of nodes, their location and the size of the network as a whole; the performance of wireless sensor networks can be optimized by applying appropriate routing protocols, which are proposed in the available literature; the proposed analyses of wireless sensor networks can be used for IoT scenarios to determine environmental parameters.

The article is structured as follows: An analysis of available literature sources has been made (Section 2). Six cities in Bulgaria have been selected (Section 3). WSN topologies have been created, depending on the regulations for each city. A numerical analysis of wireless sensor networks in terms of their energy efficiency was performed, and the obtained results are compared with those of the available literature (Section 4). Summaries and conclusions have been made (Section 5).

2. Related Works

The parameters BW, SF, different loads and modulation methods in LoRa were used by Aref et al. [13]. The authors study the influence of the data obtained from environmental parameters on the indicated characteristics of the wireless network. According to Rida et al. [14], the communication setup parameters in LoRa have a significant impact on the performance of the wireless sensor network. Such is the power of the transmitter, BW, SF, and the number of measuring points. The choice of an appropriate method of communication depends on its application and requirements. The way data are transmitted is also an important factor in communication, as Kim et al. [15] described. These authors offer a LoRa-based network with 3000 end devices. A real test for LoRa was performed and experiments were performed to find a simplified model that reduces data loss in LoRa. The authors found that the radio signal can reach distances above 1344 m with different values of SF in a combined indoor and outdoor communication environment.

Mikhaylov [16] points out that the European 868 MHz Industrial, Scientific and Medical (ISM) band is an appropriate frequency band due to the use of low frequencies for data transmission over LoRa wireless networks in European countries. The study in this article was carried out in Bulgaria and such a permitted frequency range was used.

Lukas et al. [17] offer a system for monitoring the parameters of the environment. It is based on IoT and LoRa. The main purpose of this system is to determine the characteristics of water. When using wireless sensor networks, it is necessary to take into account exactly which environmental parameters will be measured. For example, the measurement of water characteristics differs, for example, from that of air or noise in an urban environment.

Energy consumption is of great importance when using LoRa in IoT applications. According to Liang et al. [2], the closer they are to the gateway, the less energy the wireless sensor devices consume. The sensor devices are designed to use a battery or solar panels. Reducing the energy consumed will increase the life of the device and the frequency of transmission of measurement data.

The topology of the network also needs to be taken into account [18]. The "star" topology used in LoRa, in addition to its advantages, also has some limitations related to the high load on the central node, especially in networks with multiple wireless measuring stations.

According to Fernandez-Prieto et al. [19], a complete wireless sensor data collection system covers the design of the network topology, the hardware and software of the sensor nodes, the protocols and the cloud web server platform. The authors propose a comprehensive project of the IoT network for the assessment of environmental parameters in urban conditions. The stages of network design and performance evaluation can be used in the present work.

The design of a sensor network using LoRa technology is proposed by Hristov et al. [20]. The authors design the network on three levels and offer a sensor measurement system, as well as a methodology for designing wireless sensor networks in IoT. A disadvantage of this project is that the proposed methods and tools have not been tested in a real environment.

From the review of the available literature, it can be concluded that the simulation studies of LoRa wireless sensor networks are carried out "in principle", without taking into account the influence of the environment in which they are used and the exact location of sensor devices. The evaluation of actually created networks is performed after their installation. This leads to inefficient use and detection of deficiencies once the network is installed. For this reason, the shortcomings of creating a wireless sensor network are difficult to overcome. Further research is needed on the simulation analysis of wireless sensor networks suitable for use in IoT. This analysis must take into account both the characteristics and setup of the devices with network communication and the environmental factors of the space in which they will operate.

3. Material and Methods

Data for six cities in the Republic of Bulgaria were used. The data for the location of the measuring points were provided by the municipal administration of the respective city. Table 2 shows data on cities with their geographical coordinates and the number of points for measuring environmental parameters. The geographical coordinates are those of the administrative building of the respective city, which is assumed to be the location of the base station in the wireless sensor network. The cities of Shumen and Yambol have the lowest number of measuring points, and the most are in the city of Plovdiv. The distances between nodes and base station were calculated by the Euclidean distance method. The maximum distance of nodes or the radius of the wireless sensor network is specified. The shortest distance is in Burgas, and the longest distances are in Varna and Plovdiv. The average distances to the nodes are indicated. The largest is for T6 (Plowdiv) and the smallest in T1 (Burgas). The number of nodes that are close to the base station is also important. Most are in T1 (9), while in other cities there are only 1. Nearby nodes communicate more often with the base station and therefore their energy consumption is higher than that of others in the network. This increases the overall consumption of the network as a whole.

Table 2. Data for cities in Bulgaria.

Designation	City	Geographical Coordinates in WGS84		Number of Nodes	Maximum Distance, km	Average Distance, km	Nodes at $d \leq 2$ km
		°N	°E				
T1	Burgas	42.501981	27.467896	39	8	3	9
T2	Varna	43.207884	27.906593	59	12	8	1
T3	Shumen	43.270546	26.935378	15	10	8	1
T4	Yambol	42.484188	26.508795	15	11	9	1
T5	Plovdiv	42.144543	24.744737	65	12	8	1
T6	Pleven	43.408031	24.619381	36	11	10	1

Figure 1 shows maps in scale 1:75,000 and scale of 1 km of the cities in Bulgaria and the location of the measuring points (in blue). Open Street Map maps (https://www.openstreetmap.org, accessed on 1 April 2022) were used. Cities with a population of over 100,000 people were selected—Varna, Burgas, andPleven. In these cities, there are measuring networks with up to 5 points, and the other measurements were performed manually or by calculation. The smaller cities in terms of population are Shumen and Yambol. In them, the measurement of the environmental parameters was performed by manual measuring means or by calculation.

It is typical for the cities in Bulgaria to have established measuring points, regardless of the method by which the measurement data are obtained. These measurement points were used in the present work as a basis for the simulation of wireless sensor networks, with possibilities for application in IoT.

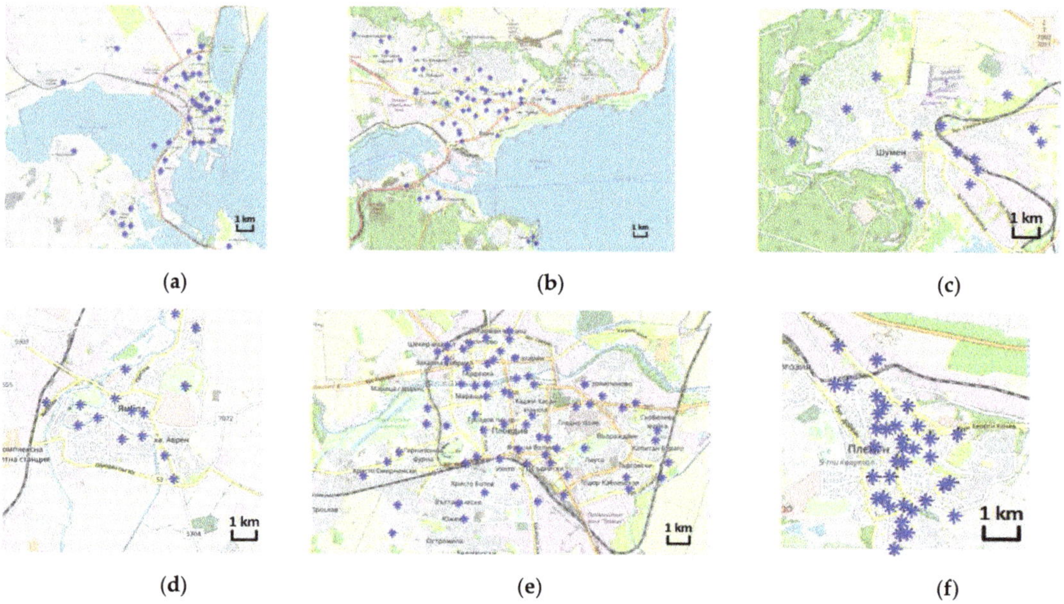

Figure 1. Location of measuring points by cities in Bulgaria. (**a**) T1 Burgas; (**b**) T2 Varna; (**c**) T3 Shumen; (**d**) T4 Yambol; (**e**) T5 Plovdiv; (**f**) T6 Pleven.

The distance for transmitting measurement data is described by the following mathematical dependence [21]:

$$P_r = \frac{P_t G_t G_r \lambda^2}{(4\pi)^2 d^2} \tag{1}$$

where P_t is the transmitted power; P_r is the received power; G_t is the gain of the transmitting antenna; G_r is the gain of the receiving antenna; λ is the wavelength; d is the distance between the receiving and transmitting parties. Euclidean distance is taken into account. This model assumes that the power received is a function of the distance to transmit and receive wireless signals. When designing a WSN for a specific city, it is necessary to take into account factors such as geographical location, the presence of buildings, trees, and the capabilities of the sensor devices used.

The average energy consumption (E_A) is determined on the basis of the arithmetic mean of the energies required for transmission, reception, sleep, and idle modes. It can be defined by the following formula:

$$E_A = \frac{1}{N} \sum_{i=1}^{N} E_i \tag{2}$$

where E_i is the energies required for transmission, reception, sleep, and idle modes of operation of the WSN; N is the number of reports for which this averaging is performed. In the present work, N is assumed to be 50 times as many as the number of rounds.

The consumed energy for 50 rounds is determined. During the operation of the sensor network, each sensor node periodically sends data to the base station. One round of the transmitted data represents the length of time it takes to send a unit of data (4000 bits) to the base station.

There is no description of all the functions of the used software product. Therefore, they are represented in the form y = f(x). The presented results of the work should be accepted within the described conditions and in those of the used software product.

Another factor that can increase the life of the network is the residual energy in each sensor node. It is calculated by a coefficient representing the current energy level of the individual sensor nodes. This remaining energy is calculated by the formula:

$$E_R = f(E_0,\ E_m) \tag{3}$$

where E_0 is the initial energies of each node; E_m is the energies required for the operation of the nodes in the WSN.

The energy consumption of the whole network (E) through all rounds is determined.

$$E = f(R) \tag{4}$$

where R is the number of rounds.

The number of received packets (N_p) for all rounds is determined.

$$N_p = f(R) \tag{5}$$

where R is the number of rounds.

Data for the LoRa Wan module operating at 868 MHz were used. The data for the base station corresponds to LoRa WAN Gateway, 868 MHz (Shenzhen Dragino technology development Co., Ltd., Shenzhen, China). Table 3 shows the main characteristics of the LoRa WAN devices.

Table 3. Data on LoRa WAN devices.

Characteristic	LoRa WAN Gateway	LoRa Wan Module
Max Output Power, dBm	27	20
Sensitivity, dBm	−140	−148
Power Input, V	12	5
bit rate up to, kbps	300	300
Transmitting power, mW	100	100
Antenna gain (Transmit), dBm	168	168
Antenna gain (Receive), dBm	27	27
Frequency, MHz	868	868

The simulation study of the wireless sensor network was performed in the Matlab 2017 b software system (MathWorks Inc., Natick, MA, USA).

A software tool was used to evaluate the energy efficiency of wireless sensor networks [22]. The performance of the six WSNs was evaluated under the following conditions: All sensor nodes have the same initial energy (E0). The location of the nodes is in accordance with the measurement points provided in the normative documents for each city. Each node transmits/receives packets when it has enough energy to do so. Receiving packets from all nodes to the base station is considered one round.

A summary analysis of the obtained results was made with the method "Principal component analysis" (PCA) [23]. This is a method of reducing the amount of input data. The aim is to find and interpret the fundamental interdependencies between the characteristics in the data set. Features that are similar are combined and transformed into new features called principal components. Before being processed by this method, the data were normalized in the range [0, 1], depending on their minimum and maximum values.

4. Results and Discussion

Figure 2 shows simulation diagrams of the WSN's access to the measuring points for the six cities in Bulgaria. A star topology was used. Most nodes that are close to the base station (up to 2 km) are in cities T1, T2, T4 and T5. The nodes are presented with blue points and green lines show the two-way communication with the base station. Some of the green lines are not visualized by the used software, but there are calculations for the nodes.

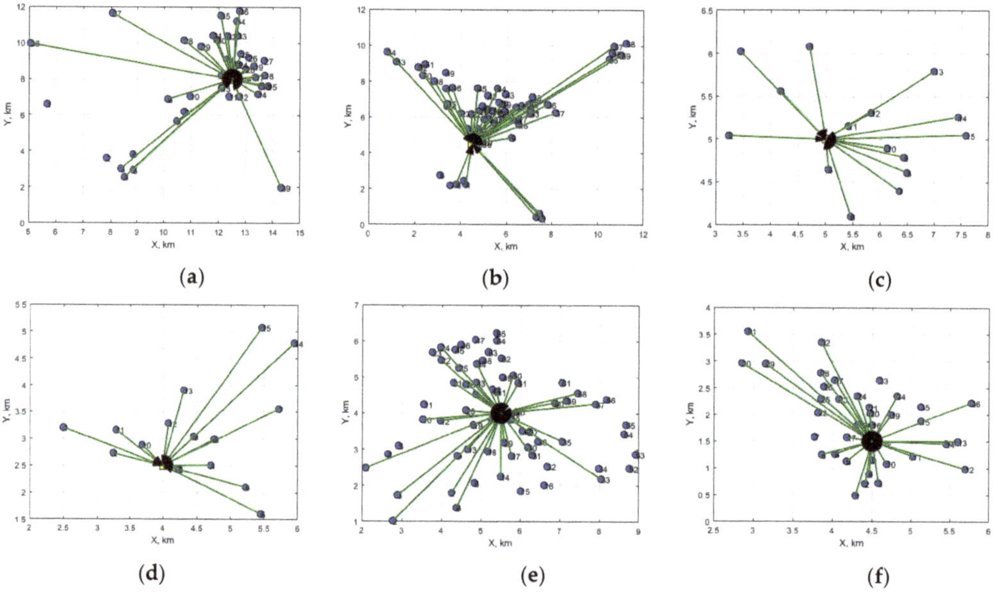

Figure 2. Simulation schemes for cities in Bulgaria. (**a**) T1 Burgas; (**b**) T2 Varna; (**c**) T3 Shumen; (**d**) T4 Yambol; (**e**) T5 Plovdiv; (**f**) T6 Pleven.

Figure 3 shows a graph of maximum energy consumption in wireless sensor networks for different cities in Bulgaria. The maximum energy consumption for the entire WSN in every city is presented. The highest energy consumption is observed in the cities T1 and T3. In T1, this is due to the many nodes that are located near the base station. In T3, the characteristics of the geographical location of the sensor nodes with network communication have a significant impact and increase the energy consumption in the network. Similar conditions are observed for the T6 network. Although T2 and T5 are significantly larger cities than the others, their average energy consumption is significantly lower compared to the networks in other cities.

Figure 3. Maximum energy consumption for WSN by cities.

Joules are used as the measurement unit because Watts measure the rate of using energy, which is not useful for reporting the amount of energy that WSNs use. We must consider that Watts were used for such a long time, which would give a figure and was a convenient way to obtain measurements; however, the actual unit for an amount of consumed energy is the Coulomb, which is a small unit, so for WSN energy consumption measurement, we need of a bigger unit. That is why Joules is the preferred measurement unit in WSN.

Figure 4 shows the average level of power consumption in wireless sensor networks, depending on the number of rounds. The results obtained during the continuous operation of the network show that in the city of T2, the distribution of nodes has a significant impact on average energy consumption. With T6, the lowest power consumption is observed, compared to other networks; this is because the nodes located close to the base station are significantly fewer and this generally reduces the energy consumption during the operation of the network. In other cases, for T1, T3, T4 and T5 networks, the average energy consumption is similar in value. Only in T1, due to the presence of a number of nodes near the base station, when transmitting more than 30 rounds, there is an increase in energy consumption. The nodes that are near the base station (less than 1 km) have greater energy consumption because they communicate more often with the station in comparison with other notes at greater distances.

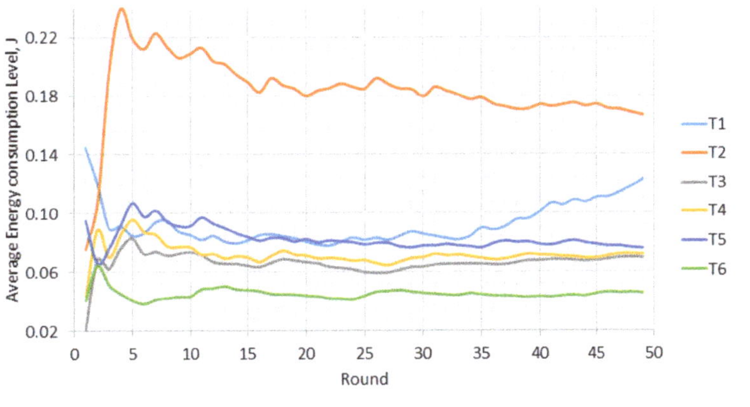

Figure 4. Average level of energy consumption for WSN by cities.

Figure 5 shows the remaining energy level in the wireless sensor networks, depending on the number of rounds. The highest levels of residual energy are observed in the T2 and T5 networks. As can be seen from the previous graphs, in these networks, the maximum and average levels of energy consumption are maintained during operation of the network. For this reason, these two networks operate more energy efficiently than the others. The lowest residual energy levels are observed in T4. Although there are fewer nodes, the distances in the city are not great (up to 4 km). This makes the location of the nodes close to the base station and therefore increases the amount of data exchanged between them. A zoomed example for T1 is given. The changes are in the range near 0.0006 J.

Figure 6 shows the level of power consumption in wireless sensor networks, depending on the number of rounds. The greater the number of nodes in the network, the greater the energy consumption. For example, T2 has 59 nodes and, compared to the others, has the highest values of energy consumption. Although some networks also have a large number of nodes, their energy consumption is lower due to the shorter transmission distances.

Figure 7 shows the number of packets received (N_p) in the wireless sensor networks in relation to the number of rounds. As can be seen from the graph, the number of packets received increases as more rounds mean more data are transmitted to the base station. With the number of rounds, the number of received packets remains high enough. It can be seen that different plots are overlapped. The six curves are close to each other, showing that for the six WSNs, there is no clear trend showing that the number of packet losses on a long path is higher than that on a short path, otherwise the nodes should receive more packets with a smaller number of rounds.

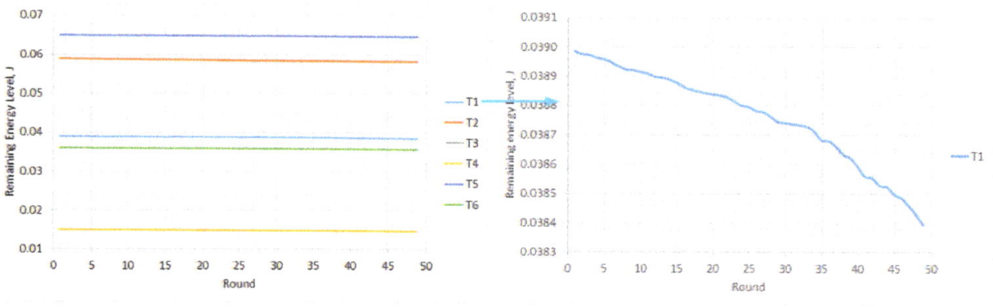

Figure 5. Remaining energy levels for WSN by cities.

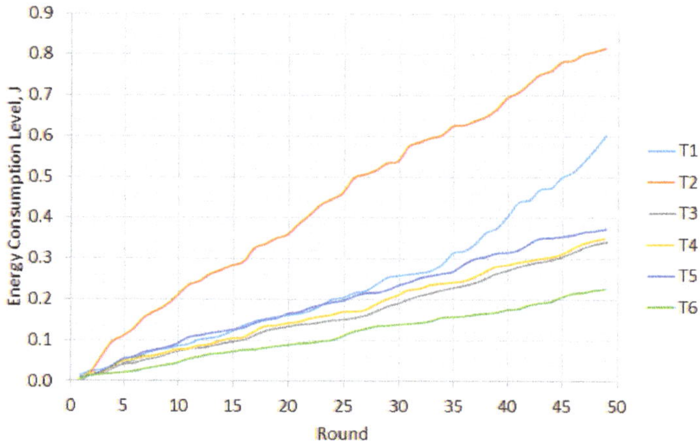

Figure 6. Energy consumption levels for WSN by cities.

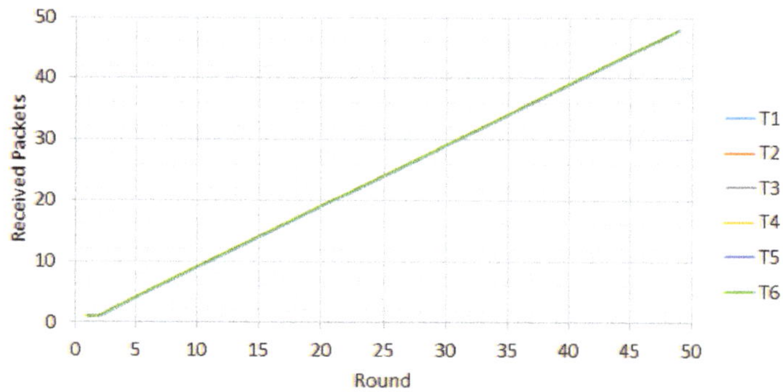

Figure 7. Number received packets of WSN by cities.

An analysis of the principle components has been made. Table 4 shows the normed values (in the interval [0, 1]) of the WSN characteristics presented above. It can be seen that the normalized value of Np in all cases is $N_p = 0$. This is because, as the graph shows, the six curves for this characteristic are close to each other. The number of columns is 6 and the number of rows is 5. By rows, the data can be reduced to a maximum of four principal components, and by columns to a maximum of five principal components.

Table 4. Normed data of WSN characteristics.

City Characteristic	T1	T2	T3	T4	T5	T6
N_p	0.00	0.00	0.00	0.00	0.00	0.00
ECL	0.28	0.75	0.12	0.14	0.20	0.00
A_r	0.37	0.67	0.00	0.00	0.77	0.33
E_a	0.26	0.75	0.11	0.14	0.21	0.00
MEC	0.60	0.00	0.20	0.07	0.00	0.07

Np—number of received packets; ECL—energy consumption level; Ar—remaining energy levels; Ea—average energy consumption; MEC—maximum energy consumption.

The first three principal components were chosen because they describe a variance in the data of over 95%. The results are shown in Figure 8. PCA analysis is used to show the cities, grouped by their specific WSN energy consumption parameters. There is a grouping of cities, depending on energy consumption in the WSN. The coastal cities T1 and T2 are in a common group, where the remaining energy level in the network has the highest values. The smaller cities T3 and T4 are in a common group and their total energy consumption is the best compared to other cities. Networks with approximately the same radii in cities T5 and T6 are in a common group and their average level of energy consumption has the best characteristics compared to other cities. The number of received packets remains in the center of the coordinate system, which shows that it is the same for all cities.

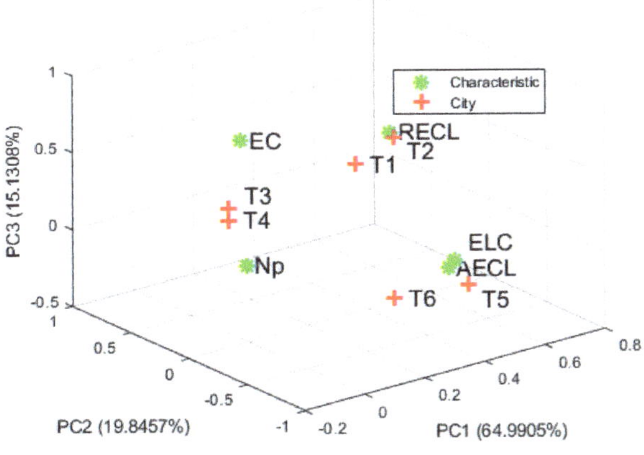

Figure 8. Principal component analysis of the WSN. EC-Energy consumption; RECL-Remaining energy consumption level; ECL-Energy consumption level; AECL-Average energy consumption level; Np-Number of received packets.

From the analysis of variants for wireless sensor networks for six cities in Bulgaria, it was found that at distances up to 2 km and a number of nodes over 30, higher values of energy consumption in the wireless sensor network are obtained. To reduce energy consumption, it is necessary to apply optimized routing protocols. Shah et al. [24] offer such a protocol/algorithm called distance-based dynamic duty-cycle allocation (DBDDCA). Through this protocol, nodes located at great distances from the base station transmit for shorter periods of time, which reduces the energy consumed in the network as a whole.

Such a routing protocol would be appropriate in the cities of Burgas (T1) and Varna (T2), because due to the specifics of the location of cities along the coast. The distances between nodes and the base stations in T1 and T2 are greater than other cities.

Ivanov's research [1] has been supplemented. It has been proven that the choice of communication modules and protocols for data transmission in wireless sensor networks depends on the specific conditions under which the measurement of environmental parameters will be performed. This thesis has been proven in six cities in Bulgaria. Machine Learning and Artificial Intelligence can be applied for that purpose [25]. A WSN data routing protocol is proposed by Arumugam et al. [26]. It depends on the optimal location of the base station. A limitation of this method is that multiple delays are available. Delays in data transmission are caused by a large number of operations in the algorithm. Software models [27] can be applied to make WSN more cost-effective.

5. Conclusions

As a result of the calculations and analyses of WSNs for six cities in Bulgaria, it was found that when designing WSNs, it is necessary to take into account the number of nodes, their location, and the size of the network as a whole.

It has been found that a greater number of nodes in the WSN leads to an increase in the average power consumed in the network. On the other hand, depending on the location of these nodes, the energy consumed may decrease.

The methodology proposed in the present paper for analysis of energy efficiency of WSN can be used in the design of wireless sensor networks for determining environmental parameters, with the possibility of application in IoT.

It can be stated that the proposed analyses of wireless sensor networks can be used for IoT scenarios. The use of results in IoT scenarios can be further improved. This should include the implementation of some security mechanisms for secure communication between the platform and the sensor nodes. Furthermore, the presented methods and tools can be used in the training of future specialists in the subject area.

Research can be continued with an analysis of the extent to which environmental measurement devices can be used in different mobile sensor network applications and how different types of transmitted data are affected by the location of sensor devices. It is also necessary to ensure the security of the transmission of measurement data. The performance of wireless sensor networks can be optimized by applying appropriate routing protocols, which are proposed in the available literature.

Author Contributions: Conceptualization, Z.Z. and T.G.; methodology, V.S.; software, Z.Z.; validation, A.T. and V.S.; formal analysis, T.G.; investigation, Z.Z. and A.T.; resources, Z.Z.; data curation, T.G.; writing—original draft preparation, Z.Z. and A.T.; writing—review and editing, V.S.; visualization, Z.Z.; supervision, T.G. All authors have read and agreed to the published version of the manuscript.

Funding: This research received no external funding.

Institutional Review Board Statement: Not applicable. (Original research of the authors).

Informed Consent Statement: Not applicable. (The research is not related to working with people).

Data Availability Statement: Data will be available on request.

Conflicts of Interest: The authors declare no conflict of interest.

References

1. Ivanov, I. Simulation analysis of wireless sensor network for determination of environmental noise. *J. Inform. Innov. Technol. (JIIT)* **2022**, *2*, 26–30.
2. Liang, R.; Zhao, L.; Wang, P. Performance Evaluations of LoRa Wireless Communication in Building Environments. *Sensors* **2020**, *20*, 3828. [CrossRef] [PubMed]

3. Europen Comission. Smart Cities: Cities Using Technological Solutions to Improve the Management and Efficiency of the Urban Environment. Available online: https://ec.europa.eu/info/eu-regional-and-urban-development/topics/cities-and-urban-development/city-initiatives/smart-cities_en (accessed on 9 March 2022).
4. Development of Updated Strategic Noise Maps for the Varna Agglomeration. SPECTRI. 2017. Available online: https://spectri.net (accessed on 24 April 2022). (In Bulgarian).
5. Lopez-Ballester, J.; Pastor-Aparicio, A.; Felici-Castell, S.; Segura-Garcia, J.; Cobos, M. Enabling real-time computation of psycho-acoustic parameters in acoustic sensors using convolutional neural networks. *IEEE Sens. J.* **2020**, *20*, 11429–11438. [CrossRef]
6. OFFNEWS. LoRa Is Already Operating in Sofia—A New Generation Network That Can Turn It into a Smart City. Available online: https://offnews.bg (accessed on 24 April 2022). (In Bulgarian).
7. Saha, D.; Shinde, M.; Thadeshwar, S. IoT based air quality monitoring system using wireless sensors deployed in public bus services. In Proceedings of the Second International Conference on Internet of things, Data and Cloud Computing (ICC'17), Cambridge, UK, 22–23 March 2017; ACM: New York, NY, USA, 2017. Article 87. pp. 1–6. [CrossRef]
8. Wang, T.; Zhang, Y.; Xiong, N.; Wan, S.; Shen, S.; Huang, S. An Effective Edge-Intelligent Service Placement Technology for 5G-and-Beyond Industrial IoT. *IEEE Trans. Ind. Inform.* **2022**, *18*, 4148–4157. [CrossRef]
9. Baker, N. ZigBee and Bluetooth: Strengths and weaknesses for industrial applications. *IEEE Comput. Control. Eng.* **2005**, *16*, 20–25. [CrossRef]
10. Lee, J.-S.; Su, Y.-W.; Shen, C.-C. A Comparative Study of Wireless Protocols: Bluetooth, UWB, ZigBee, and Wi-Fi. In Proceedings of the 33rd Annual Conference of the IEEE Industrial Electronics Society (IECON), Taipei, Taiwan, 5–8 November 2007; 8 November 2007.
11. Blobel, J.; Menne, F.; Yu, D.; Cheng, X.; Dressler, F. Low-Power and Low-Delay WLAN Using Wake-Up Receivers. *IEEE Trans. Mob. Comput.* **2022**, *21*, 1739–1750. [CrossRef]
12. Avila-Campos, P.; Astudillo-Salinas, F.; Vazquez-Rodas, A.; Araujo, A. Evaluation of LoRaWAN transmission range for wireless sensor networks in Riparian forests. In *Proceedings of MSWiM'19, Session: LPWAN and Cellular Networks*; Miami Beach, FL, USA, 25–29 November 2019, ACM digital library: New York, NY, USA, 2019; pp. 199–206.
13. Aref, M.; Sikora, A. Free space range measurements with Semtech LoRa technology. In *Wireless Systems within the Conferences on Intelligent Data Acquisition and Advanced Computing Systems: Technology and Applications (IDAACS-SWS), Proceedings of the 2014 2nd International Symposium on Wireless Systems within the Conferences on Intelligent Data Acquisition and Advanced Computing Systems, IDAACS-SWS, Odessa, Ukraine, 11-12 September 2014*; IEEE: New York, NY, USA, 2014; pp. 19–23. [CrossRef]
14. Rida, E.; Samer, L.; Melhem, E. LoRaWAN Network: Radio Propagation Models and Performance Evaluation in Various Environments in Lebanon. *IEEE Internet Things J.* **2019**, *6*, 2366–2378.
15. Kim, D.-H.; Lee, E.-K.; Kim, J. Experiencing LoRa Network Establishment on a Smart Energy Campus Testbed. *Sustainability* **2019**, *11*, 1917. [CrossRef]
16. Mikhaylov, K. On the Coverage of LPWANs: Range Evaluation and Channel Attenuation Model for LoRa Technology. In *ITS Telecommunications (ITST), Proceedings of the 2015 14th International Conference, Copenhagen, Denmark, 2–4 December 2015*; IEEE: New York, NY, USA, 2015; pp. 55–59.
17. Lukas, W.; Tanumihardja, A.; Gunawan, E. On the application of IoT: Monitoring of troughs water level using WSN. In Proceedings of the 2015 IEEE Conference on Wireless Sensor, Melaka, Malaysia, 24–26 August 2016; IEEE: New York, NY, USA, 2016; pp. 58–62.
18. Yick, J.; Mukherjee, B.; Ghosal, D. Wireless sensor network survey. *Comput. Netw.* **2008**, *52*, 2292–2330. [CrossRef]
19. Fernandez-Prieto, J.-A.; Cañada-Bago, J.; Gadeo-Martos, M.-A. Wireless Acoustic Sensor Nodes for Noise Monitoring in the City of Linares (Jaén). *Sensors* **2020**, *20*, 124. [CrossRef] [PubMed]
20. Hristov, G.; Zahariev, P.; Raychev, J.; Kinaneva, D.; Mihov, I. A methodology for environmental and air quality monitoring using LoRaWAN sensor platforms. *Proc. Univ. Ruse* **2018**, *57*, 73–82.
21. LoRaWAN Part 1: How to Get 15 km Wireless and 10-Year Battery Life for IoT. Available online: https://www.digikey.com (accessed on 15 February 2022).
22. WSNsimulatorMatlab. Available online: https://github.com (accessed on 15 February 2022).
23. Mladenov, M. Model-based approach for assessment of freshness and safety of meat and dairy products using a simple method for hyperspectral analysis. *J. Food Nutr. Res.* **2020**, *59*, 8–119.
24. Shah, I.; Maity, T.; Dohare, Y. Algorithm for energy consumption minimisation in wireless sensor network. *IET Commun.* **2020**, *14*, 1301–1310. [CrossRef]
25. Brites, I.; da Silva, L.; Barbosa, J.; Rigo, S.; Correia, S.; Leithardt, V. Machine Learning and IoT Applied to Cardiovascular Diseases Identification through Heart Sounds: A Literature Review. *Informatics* **2021**, *8*, 73. [CrossRef]
26. Arumugam, G. EE-LEACH: Development of energy-efficient LEACH Protocol for data gathering in WSN. *EURASIP J. Wirel. Commun. Netw.* **2015**, *1*, 1–9. [CrossRef]
27. Pinheiro, E.; Correia, S. Software model for a low-cost, IoT oriented energy monitoring platform. *SSRG Int. J. Comput. Sci. Eng. (SSRG-IJCSE)* **2018**, *5*, 1–5. [CrossRef]

computers

MDPI

Article

IoT Security Mechanisms in the Example of BLE

Evgeny Kalinin *, Danila Belyakov, Dmitry Bragin and Anton Konev

Faculty of Security, Tomsk State University of Control Systems and Radioelectronics, 634000 Tomsk, Russia; bds2@csp.tusur.ru (D.B.); bds@csp.tusur.ru (D.B.); kaa1@keva.tusur.ru (A.K.)
* Correspondence: kot60068@vtomske.ru

Abstract: In recent years, a lot of IoT devices, wireless sensors, and smart things contain information that must be transmitted to the server for further processing. Due to the distance between devices, battery power, and the possibility of sudden device failure, the network that connects the devices must be scalable, energy efficient, and flexible. Particular attention must be paid to the protection of the transmitted data. The Bluetooth mesh was chosen as such a network. This network is built on top of Bluetooth Low-Energy devices, which are widespread in the market and whose radio modules are available from several manufacturers. This paper presents an overview of security mechanisms for the Bluetooth mesh network. This network provides encryption at two layers: network and upper transport layers, which increases the level of data security. The network uses sequence numbers for each message to protect against replay attacks. The introduction of devices into the network is provided with an encryption key, and the out-of-band (OOB) mechanism is also supported. At the moment, a comparison has been made between attacks and defense mechanisms that overlap these attacks. The article also suggested ways to improve network resiliency.

Keywords: Bluetooth mesh; BLE; security; IoT

Citation: Kalinin, E.; Belyakov, D.; Bragin, D.; Konev A. IoT Security Mechanisms in the Example of BLE. *Computers* **2021**, *10*, 162. https://doi.org/10.3390/computers10120162

Academic Editor: Sergio Correia

Received: 29 October 2021
Accepted: 26 November 2021
Published: 29 November 2021

1. Introduction

Mesh network–network topology dynamically establishes the maximum possible number of connections between devices for efficient and resilient data transmission [1]. Usually, mesh networks are based on several wireless technologies, such as Wi-Fi, Bluetooth, Thread, and Zigbee [2–5]. However, a mesh network based on Bluetooth Low Energy (BLE) is a popular solution [3,6] due to its high popularity, low cost, and low power consumption [3,7].

The main problem of building mesh networks is to ensure the security of data transmission [8]. Since the network can cover a large area [4], it is necessary to provide efficient protection against unauthorized access by intruders.

The aim of this paper is to compare wireless attacks and defense mechanisms implemented in the Bluetooth mesh standard, which was created by the Bluetooth Special Interest Group (SIG) and implemented on Bluetooth Low-Energy devices starting with 4.0 [9]. The current version of Mesh Profile 1.0.1 can be implemented on BLE devices from 5.0 and later. [10]. Furthermore, one of the tasks is to search for network vulnerabilities and provide recommendations for their elimination.

2. Wireless Network Vulnerabilities

Before considering the existing information security mechanisms, its necessary to introduce a classification of possible network attacks on wireless networks [11–13]:

- Denial of Service. The purpose of this attack is to overload the device with redundant packets, which make the device unusable [14,15];
- Eavesdropping. An attacker eavesdrops on a data exchange in order to extract useful information;

- Man In The Middle, MITM. A malicious device secretly establishes a connection between two devices and making them think they are exchanging data with each other;
- Replay Attack. A previously sent valid message captured by an intruder can be used to exploit the system functionality without an authentication procedure [15].
- Relay Attack. A malicious device establishes communication between two nodes and transmits unmodified data between them [14].

Thus, in order to ensure the information security of wireless devices, it is necessary to take these attacks into account when developing security algorithms. However, these attacks are network attacks and do not include physical interaction attacks on the device, such as attacks to obtain security parameters [16].

3. Bluetooth Mesh Overview

First of all, we need to define the terms used in the Bluetooth mesh specifications [10]. Devices that can transmit or receive messages are called nodes or provisioned devices. Devices that are not a part of any mesh network are called unprovisioned devices. The provisioning process [17] is carried out by a provisioner device, which authenticates unprovisioned devices, assigns an address, and transmits encryption keys [18]. Further configuration of a node is performed by a configuration client, which can be a part of a provisioner device or a part of another node. A configuration client transmits an application key, additional network keys, and configures subscription and publish address for each model.

Each mesh node must have at least one element. An element is an addressable entity that contains models, which defines node functionality. All data exchanges are carried out using messages, which are defined by the opcode, associated parameters, and behavior. Messages are operating with states that represent the position of an element.

There are three types of models [19]: The Server model contains one or several states, divided between one or several elements, as well as messages and behaviors associated with receiving and sending data; the Client Model contains messages necessary for requesting, changing, or using corresponding states of the server; the Control model contains the functionality of both models.

The Bluetooth specification [9] also defines two types of messages: Control Messages for controlling network operations and Access Messages for data distribution. A data exchange between nodes is defined by subscription and publication methods [20,21]. Addresses can be unicast, group, and virtual. Unicast addresses are assigned to each element.

Bluetooth Mesh specification [10] describes the different types of nodes [22]:

- A relay node is a device designed for data transmission within a mesh network. The message forwarding distance is limited by the TTL value;
- A low-power node is a device that spends most of its time in sleep mode. After waking up, a low-power device receives data from a friendly host;
- A friend node is a device that stores data for the Low-Profile node;
- A proxy node is a device that can work with BLE devices via GATT;
- A provisioner node is a device designed to register nodes in the mesh network and to distribute security keys. It also can be a configurator device.

Unlike other protocols (such as Zigbee, Thread, etc.), which are based on the use of routed networks, Bluetooth Mesh uses managed flooding for data transfer within the network [19,23,24]. Managed flooding is a node organizing method within the mesh network, which is a compromise between packets routing and uncontrolled forwarding of received packets. The essence of the method is that incoming packages have a time-to-live (TTL) value, which decreases with each packet transmission until the TTL value becomes zero. Furthermore, all incoming packets are cached to avoid re-forwarding. However, there is an implementation of routed networks based on BLE technologies [25,26].

Bluetooth Mesh is based on Bluetooth Low-Energy technology [10], which works in the 2.4 GHz frequency range. The entire frequency range is divided into 40 channels of

2 MHz each [9,27,28]. Since Bluetooth Mesh usually works without active connections, all communication happening in advertising channels by sending a special non-scanned type of advertising message "ADV_NONCONN_IND" [29]. Interaction between nodes occurs through one of three advertising channels: 37, 38, 39 [14,21].

3.1. Bluetooth Mesh Layers

Bluetooth Mesh has a layered model based on Bluetooth Low Energy. The Bluetooth mesh architecture consists of the following layers [20,21,28,30]:

- The models layer defines model implementation, its behavior, state;
- The foundation models layers defines network configuration and models management;
- The access layers defines application interaction with the upper transport layer to determine data format, data encryption process, and data verification;
- The upper transport layer defines message encryption and verification methods by using the application key generated for each device;
- The lower transport layer performs segmentation of transmitted messages and assembly of incoming messages;
- The network layer defines a message format for transfer data across network elements through a data link layer and message encryption. It also manages messages to be relayed, accepted, or rejected;
- The bearer layer defines packets handling methods such as transmitting data into advertising bearer, which is used in scanning and advertising state. Data transmitting through GATT, which allows communication with regular BLE devices using proxy nodes.

For security reasons, Bluetooth Mesh uses two types of keys for data transmission [28]: AppKey using for data encryption on the upper transport layer, and NetKey for data encryption on the network layer. The same NetKey is used for all nodes within the same network. Key separation allows an intermediate node to verify message integrity and forward it without exposing content, which protects data from unauthorised access. In addition, key separation allows you to secure not only from an eavesdropping attack but also from a relay attack, in which the node that is part of the network can read the messages that are not intended for it [29].

In addition to using AppKey and NetKey, there is a unique key for each device called DevKey. This key is known only to the device itself and the configuration client. DevKey is used for secure communication between the node and the configuration client. Just like the application key, the device key is used in the upper transport layer.

3.2. Authentication and Encryption

Message exchange is secured using the AES-CCM algorithm [10,31]. On the upper transport layer used an application key called AppKey for data encryption and authentication. On the network layer used two security keys: the EncryptionKey for data encryption and authentication and the Privacy key for obfuscation of the message headers. The EncryptionKey and PrivacyKey are divided from NetKey.

Due to the use of the AES-CCM algorithm, all messages have an authentication tag called the message integrity check (MIC).

A unique number that can only be used once (Nonce) is used to encrypt data. Using Nonce protects against replay attacks. Bluetooth Mesh defines four types of Nonce: network nonce (Figure 1), application nonce (Figure 2), device nonce, and proxy nonce.

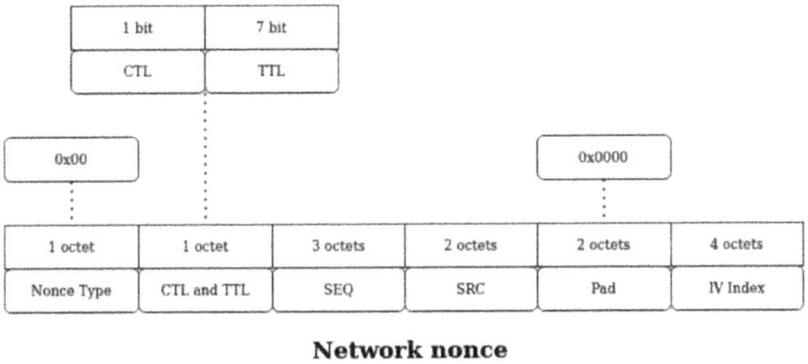

Figure 1. Network Nonce.

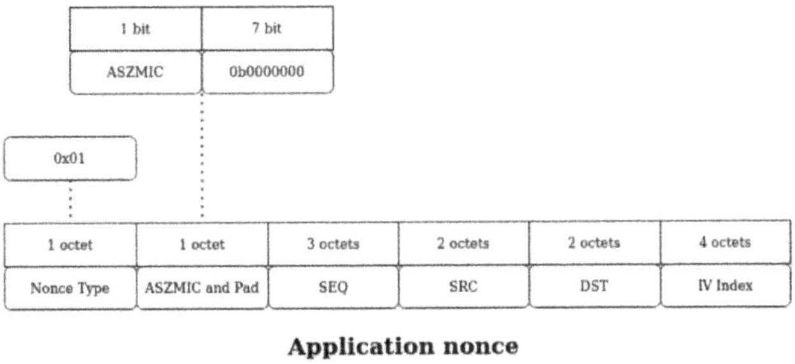

Figure 2. Application Nonce.

A nonce value contains a field type, sequence number (SEQ), source address (SRC), destination address (DST), an initialization vector (IV Index), flag type, TTL value, and ASZMIC flag to indicate the segmented message.

Message headers are obfuscated to hide identifying information, such as source address (SRC), sequence number (SEQ), etc. Header obfuscation provides protection against eavesdropping attacks. The obfuscation process using the AES algorithm with PrivacyKey and an encrypted network message.

Figure 3 presents the generation of a secure network message. The generated network message contains public information about the IVI value, the least significant bit of the IV Index, and network identifier (NID), which is used to determine the encryption key. Network identifier derived from NetKey, EncryptionKey, and PrivacyKey.

Transmitted messages can be eavesdropped on and resent later in an unmodified form. This attack is a Replay attack, in which the same message is transmitted several times, which has a malicious impact on the network. To secure the network from that kind of attack, Bluetooth Mesh defines the sequence number value (SEQ). Nodes increments a sequence value for each message transmission. If a node receives a message with a sequence number lower than previous messages, it will be rejected.

If the sequence number (SEQ) reaches its maximum value, the IV Index update procedure will be started. The IV Index is an equal value for all nodes within a network. The IV Index is updated periodically to avoid reuse, and the update procedure can be launch by any node. The IV Index must be equal to or greater than the value of the next message.

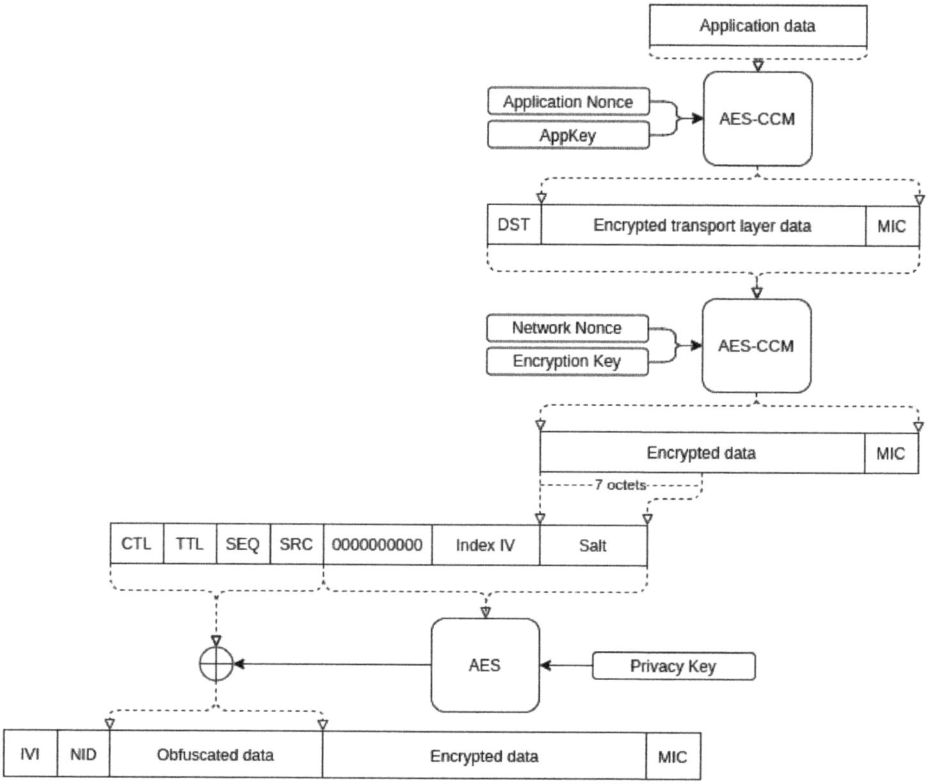

Figure 3. Formation of a secure network message.

3.3. Provisioning Procedure

A mesh network can be created by a provisioner device, which searches for unprovisioned devices and registers them on the network. The provisioner protects the network against using malicious unprovisioned devices.

The provisioning process begins after the unprovisioned device broadcasts advertising messages saying that it is available. When the provision finds the message data, it sends a provision request to the unprovisioned device and information about the supported security algorithms, public key, etc. After the response, the public key is exchanged between the provisioner and the unprovisioned device. Using the Elliptic Curve Diffie–Hellman (ECDH) protocol [32] during the provisioning process, the provisioner sends encrypted security parameters. To prevent an attack man-in-the-middle (MITM), an optional Out-of-Band (OOB) mode [33] can be used (for example, passphrase input, use of NFC technology, etc.). Once a secure connection has been established, the devices exchange provisioning data, such as NetKey, AppKey, Index IV, TTL value, node address, etc.

Discarded nodes must be disposed of, so keys stored within the device cannot be used for attacks on the network. In order to protect against such an attack, the provisioner device adds a disposed-of node to the blacklist, and each node begins the security key update procedure. The provisioner provides new NetKey and AppKey for each node, except devices from the blacklist.

4. Experiment

Table 1 shows a summary comparison of network attacks with Bluetooth Mesh protection methods. As shown, Bluetooth Mesh is vulnerable to denial of service attacks.

Since communications with nodes within a network happen by using only three channels, in an environment with high radio frequency interference, it can cause a significant loss of data. For example, general BLE devices use these channels to notify other devices, and the same channels are used by beacons to continuously broadcast messages [3,5]. Furthermore, the same frequency range is used by other technologies, such as WiFi, ZigBee, etc. [8,23].

Table 1. Comparison of network attacks and protection methods.

Attack Type	Protection Methods
Denial of Service	-
Eavesdropping	Message encryption and message header obfuscation
Man In The Middle	Authentication during node provisioning process by using OOB
Replay Attack	Use of the sequential number (SEQ) and IV Index
Relay Attack	Authentication during node provisioning process by using OOB and using two-level key separation

We have experimentally demonstrated the possibility of denying a service attack on a Bluetooth Mesh network, which continuously transmits sequentially numbered messages to identify lost packets. For this purpose, our experimental setup includes two developer boards nRF52840 [34] from Nordic Semiconductor working within the Bluetooth Mesh network; these devices are also featured in [35] as a low-cost test bed, three SDR ADALM-PLUTO [36] from Analog Devices, which all generate radio frequency interference on the same three channels. The first node sends sequence numbers, and the second node receives them and transmits them to the computer. The ADALM-PLUTO SDR was chosen because it can generate or acquire RF analog signals in a range from 325 MHz to 3800 MHz, and BLE work on 2.4 GHz. The ADALM-PLUTO SDR was handled with GNURadio software in order to control and generate an interference at a specific frequency.

The experiment was carried out in the research laboratory of the Tomsk State University of Control Systems and Radioelectronics. Current research at the time included various wireless devices, operating at 802.11, and BLE standards at the same 2.4 GHz frequency band. This is conducted in order to create as realistic an environment as possible with possible radio interference.

The result is shown in Figure 4, which shows the arrival time of the packet and the channel on which the packet was received. The experiment consisted of five stages, 10 min each: in the first stage, the jammers were turned off; in the second stage, channel 37 was muted; in the third stage, 37 and 38 channels were muted; in the fourth stage, 37, 38, and 39 channels were muted; in the fifth stage 38 and 39 channels were muted. As a result, in the fourth stage, not all packets were drowned; this is explained by the close location of the mesh network nodes, as well as the low power of the jammers. The experiment showed that this network does not have any algorithms for protection against jammer attacks, such as calculating the location of jammers, which are discussed in the article by Dhivyasri et al. [37].

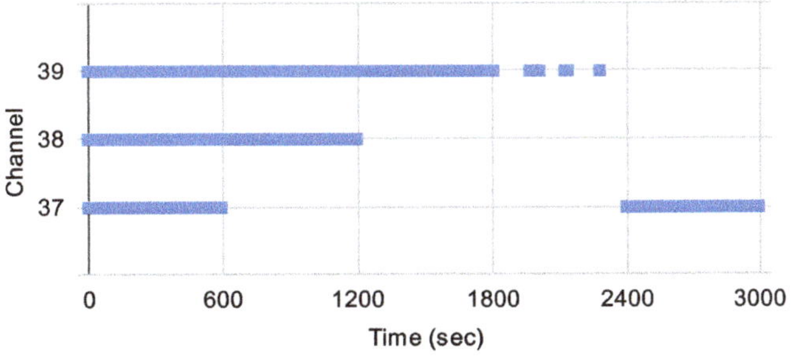

Figure 4. Received packets for each channel.

The following figures show the percentage of packet loss: the percentage of packet loss for each stage is shown in Figure 5, and the dynamics of changes in the percentage of total packet loss are shown in Figure 6.

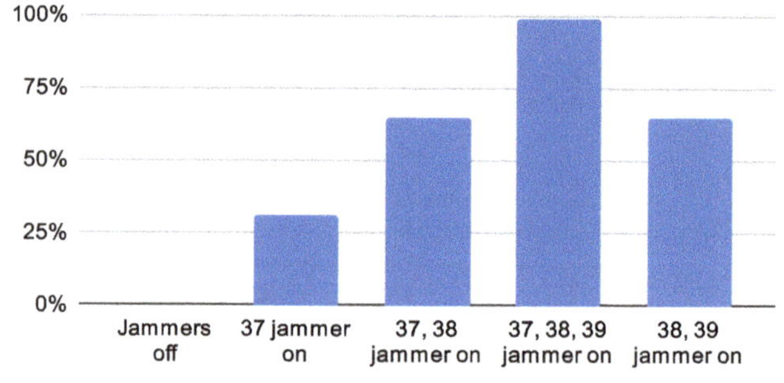

Figure 5. The percentage of packet loss for each stage.

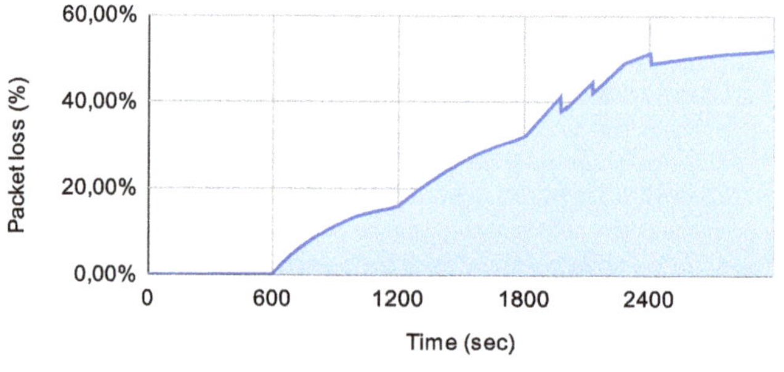

Figure 6. The dynamics of changes in the percentage of total packet loss.

The experiment confirms that with a small amount of equipment, it is possible to disrupt the network. Therefore, packet loss can be critical for systems, such as healthcare systems used to monitor patients' vital signs and their location [38]. The solution is to increase the number of used channels for data transmission, and it is necessary to use all available 40 channels. This can increase the fault tolerance of the network and make it difficult to implement such an attack.

5. Conclusions

The cheapening of microchip production has led to the rapid growth of the Internet of Things and the proliferation of various sensors and sensors. However, due to the large number of devices, it has become more difficult to deploy a wired network to provide communication between different devices. The solution to this problem is to build a wireless network. A wireless network can be deployed without any installation work, and it can exchange data between multiple devices and span large areas. However, a wireless network also has disadvantages. Due to the fact that data in a wireless network are transmitted in a shared radio environment, a collision can occur, an attacker can eavesdrop or send modified data packets, as well as jam data transmission. As a consequence, it is necessary to guarantee the protection of the transmitted data at a high level, as well as to ensure the possibility of data transmission in the presence of interference.

As a result of the work conducted, the main mechanisms of Bluetooth Mesh protection were considered, possible types of attacks on the wireless network were given, and a summary table was compiled reflecting weaknesses in network protection. Based on the table, an attack vector was found. An experiment has been successfully conducted demonstrating that despite the protection measures provided in the Bluetooth Mesh standard, it is possible to carry out a simple attack that paralyzes the operation of the entire distributed network. To increase the complexity of the attack presented in the experiment, it is necessary for the network to provide data transmission on all available channels and not only on advertising channels, which are quite noisy due to the prevalence of BLE devices and can be easily blocked by three jammers.

In the future, we plan to delve deeper into the security algorithms of the Bluetooth Mesh, as well as implement protection against the attack demonstrated in the experiment.

Author Contributions: Conceptualization, D.B. (Danila Belyakov) and E.K.; methodology, D.B. (Danila Belyakov) and E.K.; software, D.B. (Danila Belyakov) and E.K.; validation, E.K., D.B. (Dmitry Bragin), and A.K.; formal analysis, E.K.; investigation, E.K.; resources, D.B. (Dmitry Bragin) and A.K.; data curation, E.K.; writing—original draft preparation, D.B. (Danila Belyakov) and E.K.; writing—review and editing, D.B. (Danila Belyakov); visualization, D.B. (Danila Belyakov); supervision, D.B. (Dmitry Bragin); project administration, D.B. (Dmitry Bragin); funding acquisition, D.B. (Dmitry Bragin). All authors have read and agreed to the published version of the manuscript.

Funding: The article was prepared as part of the implantation of the «Leading research center (LRC) «Trusted Sensor Systems», financial support provided by the Ministry of Digital Development, Communications and Mass Media of the Russian Federation and Russian Venture Company (RVC JSC) (Agreement №009/20 dated 4 October 2020).

Institutional Review Board Statement: Not applicable.

Informed Consent Statement: Not applicable.

Data Availability Statement: Not applicable.

Conflicts of Interest: The authors declare no conflict of interest.

References

1. Cardullo, P.; Roio, D. *Mesh Networks*; Wiley Online Library: Hoboken, NJ, USA, 2020; pp. 1–3. [CrossRef]
2. Cilfone, A.; Davoli, L.; Belli, L.; Ferrari, G. Wireless Mesh Networking: An IoT-Oriented Perspective Survey on Relevant Technologies. *Future Internet* **2019**, *11*, 99. [CrossRef]
3. Zanaj, E.; Caso, G.; De Nardis, L.; Mohammadpour, A.; Alay, O.; Di Benedetto, M.G. Energy Efficiency in Short and Wide-Area IoT Technologies—A Survey. *Technologies* **2021**, *9*, 22. [CrossRef]
4. Basu, S.; Baert, M.; Hoebeke, J. QoS Enabled Heterogeneous BLE Mesh Networks. *J. Sens. Actuator Netw.* **2021**, *10*, 24. [CrossRef]
5. Lin, Y.W.; Lin, C.Y. Beyond Beacons—An Interactive Positioning and Tracking System Solely Based on BLE Mesh Network. 2018, pp. 351–362. Available online: https://link.springer.com/chapter/10.1007/978-3-319-65521-5_30 (accessed on 28 June 2021).
6. Dhandapani, K.; Harshavardhan, A.; Kumar, V.; Sunitha, D.; Korra, S. BLE in IoT: Improved link stability and energy conservation using fuzzy approach for smart homes automation. *Mater. Today Proc.* **2021**, in press. [CrossRef]
7. Darroudi, S.M.; Caldera-Sànchez, R.; Gomez, C. Bluetooth Mesh Energy Consumption: A Model. *Sensors* **2019**, *19*, 1238. [CrossRef] [PubMed]
8. Angelov, K.; Sadinov, S.; Kogias, P. Deployment of mesh network in an indoor scenario for application in IoT communications. *IOP Conf. Ser. Mater. Sci. Eng.* **2021**, *1032*, 012004. [CrossRef]
9. SIG, B. Core Specification 5.3. Available online: https://www.bluetooth.com/specifications/specs/core-specification/ (accessed on 1 July 2021).
10. SIG, B. Mesh Profile 1.0.1. Available online: https://www.bluetooth.com/specifications/specs/mesh-profile-1-0-1/ (accessed on 1 July 2021).
11. Sen, J. *Security and Privacy Issues in Wireless Mesh Networks: A Survey*; Springer: Berlin/Heidelberg, Germany, 2013; pp. 189–272. [CrossRef]
12. Ghori, M.; Wan, T.C.; Anbar, M.; Sodhy, G.; Rizwan, A. Review on security in bluetooth low energy mesh network in correlation with wireless mesh network security. In Proceedings of the 2019 IEEE Student Conference on Research and Development (SCOReD), Bandar Seri Iskandar, Malaysia, 15–17 October 2019; pp. 219–224. [CrossRef]
13. Toçilla, A. Overview of Security in Wireless Mesh Networks (WMNs). 2020. Available online: https://www.researchgate.net/publication/341043358_Overview_of_Security_in_Wireless_Mesh_Networks_WMNs (accessed on 28 June 2021).
14. Sevier, S.; Tekeoglu, A. Analyzing the Security of Bluetooth Low Energy. In Proceedings of the 2019 International Conference on Electronics, Information, and Communication (ICEIC), Auckland, New Zealand, 22–25 Jane 2019; pp. 1–5. [CrossRef]
15. Gupta, D.S.; Wani, A.; Kumar, S.; Srivastava, A.; Sharma, D. *Wireless Mesh Network Security, Architecture, and Protocols*; IGI Global: Hershey, PA, USA, 2019; pp. 1–27. [CrossRef]

16. Adomnicai, A.; Fournier, J.; Masson, L. Hardware Security Threats Against Bluetooth Mesh Networks. In Proceedings of the 2018 IEEE Conference on Communications and Network Security (CNS), Beijing, China, 30 May–1 June 2018; pp. 1–9. [CrossRef]
17. Hortelano, D.; Olivares, T.; Ruiz, M.C. Providing interoperability in Bluetooth mesh with an improved provisioning protocol. *Wirel. Netw.* **2021**, *27*, 1–23. [CrossRef]
18. Ghori, M.; Wan, T.C.; Sodhy, G. Bluetooth Low Energy Mesh Networks: Survey of Communication and Security Protocols. *Sensors* **2020**, *20*, 3590. [CrossRef] [PubMed]
19. SIG, B. Mesh Model 1.0.1. Available online: https://www.bluetooth.com/specifications/specs/mesh-model-1-0-1/ (accessed on 1 July 2021).
20. Suthar, F.A.; Patel, R.K.; Prajapati, J.B. Overview of Wireless Mesh Network's in Bluetooth Mesh. 2019. Available online: https://ssrn.com/abstract=3817363 (accessed on 1 July 2021).
21. Baert, M.; Rossey, J.; Shahid, A.; Hoebeke, J. The Bluetooth Mesh Standard: An Overview and Experimental Evaluation. *Sensors* **2018**, *18*, 2409. [CrossRef] [PubMed]
22. Wan, Q.; Liu, J. Smart-Home Architecture Based on Bluetooth mesh Technology. *IOP Conf. Ser. Mater. Sci. Eng.* **2018**, *322*, 072004. [CrossRef]
23. Brandao, A.; Lima, M.; Abbas, C.; García Villalba, L. An Energy Balanced Flooding Algorithm for a BLE Mesh Network. *IEEE Access* **2020**, *8*, 97946–97958. [CrossRef]
24. Darroudi, S.M.; Gomez, C. Bluetooth Low Energy Mesh Networks: A Survey. *Sensors* **2017**, *17*, 1467. [CrossRef] [PubMed]
25. Sirur, S.; Juturu, P.; Gupta, H.P.; Serikar, P.R.; Reddy, Y.K.; Barak, S.; Kim, B. A mesh network for mobile devices using Bluetooth low energy. In Proceedings of the 2015 IEEE SENSORS, Busan, Korea, 1–4 November 2015; pp. 1–4. [CrossRef]
26. Murillo, Y.; Reynders, B.; Chiumento, A.; Malik, S.; Crombez, P.; Pollin, S. Bluetooth now or low energy: Should BLE mesh become a flooding or onnection oriented network? In Proceedings of the 2017 IEEE 28th Annual International Symposium on Personal, Indoor, and Mobile Radio Communications (PIMRC), Montreal, QC, Canada, 8–13 October 2017; pp. 1–6. [CrossRef]
27. Zhang, Y.; Weng, J.; Dey, R.; Fu, X. Bluetooth Low Energy (BLE) Security and Privacy. 2019. Available online: http://www.cs.ucf.edu/~rajib/BLE-Encyclopedia2019.pdf (accessed on 28 June 2021).
28. Darroudi, S.M.; Gomez, C.; Crowcroft, J. Bluetooth Low Energy Mesh Networks: A Standards Perspective. *IEEE Commun. Mag.* **2020**, *58*, 95–101. [CrossRef]
29. Hernandez-Solana, A.; Pérez Díaz de Cerio, D.; Garcia-Lozano, M.; Valdovinos, A.; Valenzuela, J.L. Bluetooth Mesh Analysis, Issues and Challenges. *IEEE Access* **2020**, *8*, 53784–53800. [CrossRef]
30. Veiga, A.; Abbas, C. Proposal and Application of Bluetooth Mesh Profile for Smart Cities' Services. *Smart Cities* **2018**, *2*, 1–19. [CrossRef]
31. Padgette, J.; Bahr, J.; Batra, M.; Holtmann, M.; Smithbey, R.; Chen, L.; Scarfone, K. *NIST Special Publication 800-121 Revision 2, Guide to Bluetooth Security*; Technical Report; National Institute of Standards and Technology (NIST): Gaithersburg, MD, USA, 2017. [CrossRef]
32. Diffie, W.; Hellman, M. New directions in cryptography. *IEEE Trans. Inf. Theory* **1976**, *22*, 644–654. [CrossRef]
33. Ren, K. Provisioning a Bluetooth Mesh Network Part 2. 2017. Available online: https://www.bluetooth.com/blog/provisioning-a-bluetooth-mesh-network-part-2/ (accessed on 1 July 2021).
34. Nordic. nRF52840 DK. Available online: https://www.nordicsemi.com/Products/Development-hardware/nRF52840-DK (accessed on 1 July 2021).
35. Murillo, Y.; Reynders, B.; Chiumento, A.; Pollin, S. A Multiprotocol Low-Cost Automated Testbed for BLE Mesh. *IEEE Commun. Mag.* **2019**, *57*, 76–83. [CrossRef]
36. PlutoSDR. ADALM-PLUTO Overview. Available online: https://plutosdr.org/adalm-pluto-overview/ (accessed on 1 July 2021).
37. Dhivyasri, K. Wireless Sensor Network Jammer Attack: A Detailed Review. *Int. J. Res. Appl. Sci. Eng. Technol.* **2020**, *8*, 233–238. [CrossRef]
38. Di Marco, P.; Park, P.; Pratesi, M.; Santucci, F. A Bluetooth-Based Architecture for Contact Tracing in Healthcare Facilities. *J. Sens. Actuator Networks* **2020**, *10*, 2. [CrossRef]

computers

MDPI

Article

Enhancing Robots Navigation in Internet of Things Indoor Systems

Yahya Tashtoush [1,*], Israa Haj-Mahmoud [1], Omar Darwish [2,*], Majdi Maabreh [3], Belal Alsinglawi [4], Mahmoud Elkhodr [5] and Nasser Alsaedi [6]

1. Computer Science Department, Jordan University of Science and Technology, Irbid 22110, Jordan; israa_haj_mahmoud@yahoo.com
2. Information Security and Applied Computing Department, Eastern Michigan University, Ypsilanti, MI 48197, USA
3. Department of Information Technology, Faculty of Prince Al-Hussein Bin Abdallah II For Information Technology, The Hashemite University, P.O. Box 330127, Zarqa 13133, Jordan; majdi@hu.edu.jo
4. Computer Data and Mathematical Sciences, Western Sydney University, Sydeney, NSW 2116, Australia; b.alsinglawi@westernsydney.edu.au
5. School of Engineering and Technology, Central Queensland University, Rockhampton, QLD 4701, Australia; m.elkhodr@cqu.edu.au
6. Computer Science Department, Taibah University, Medina 2003, Saudi Arabia; nsaede@taibahu.edu.sa

* Correspondence: yahya-t@just.edu.jo (Y.T.); odarwish@emich.edu (O.D.)

Citation: Tashtoush, Y.; Haj-Mahmoud, I.; Darwish, O.; Maabreh, M.; Alsinglawi, B.; Elkhodr, M.; Alsaedi, N. Enhancing Robots Navigation in Internet of Things Indoor Systems. *Computers* **2021**, *10*, 153. https://doi.org/10.3390/computers10110153

Academic Editor: Sergio Correia

Received: 24 September 2021
Accepted: 8 November 2021
Published: 15 November 2021

Abstract: In this study, an effective local minima detection and definition algorithm is introduced for a mobile robot navigating through unknown static environments. Furthermore, five approaches are presented and compared with the popular approach wall-following to pull the robot out of the local minima enclosure namely; Random Virtual Target, Reflected Virtual Target, Global Path Backtracking, Half Path Backtracking, and Local Path Backtracking. The proposed approaches mainly depend on changing the target location temporarily to avoid the original target's attraction force effect on the robot. Moreover, to avoid getting trapped in the same location, a virtual obstacle is placed to cover the local minima enclosure. To include the most common shapes of deadlock situations, the proposed approaches were evaluated in four different environments; V-shaped, double U-shaped, C-shaped, and cluttered environments. The results reveal that the robot, using any of the proposed approaches, requires fewer steps to reach the destination, ranging from 59 to 73 m on average, as opposed to the wall-following strategy, which requires an average of 732 m. On average, the robot with a constant speed and reflected virtual target approach takes 103 s, whereas the identical robot with a wall-following approach takes 907 s to complete the tasks. Using a fuzzy-speed robot, the duration for the wall-following approach is greatly reduced to 507 s, while the reflected virtual target may only need up to 20% of that time. More results and detailed comparisons are embedded in the subsequent sections.

Keywords: local minima; target switching; trap situation; mobile robot navigation; infinite loop

1. Introduction

The number of robots deployed in the manufacturing industry has increased drastically in recent times [1]. The Internet of Things (IoT) enables autonomous and mobile robots to interact with their surroundings, sense obstacles, navigate through defiance patterns, perform a certain task, and be involved in many autonomous interactions [2]. Robots are believed to be the key enablers of industry 5.0, especially in manufacturing. When a human user initiates a task, robots that observe the process using visionary sensor devices, such as the use of a mounted camera, can then aid the workers within the production space. Robotic applications are increasingly being adopted in healthcare systems as well. The COVID-19 pandemic, for instance, has seen the use of Robots for the purpose of collecting samples from patients. Robots were also used to disinfect common spaces with larger traffic, such as in hospital entrances and supermarkets [3]. Robots are envisioned to be the key enablers for personalized healthcare systems providing assistance to patients and the elderly [4,5].

Within industrial spaces, in unmanned aerial vehicle (UAV) applications, commonly known as a drone, cameras play an important role in observing the environment the drone may encounter—particularly in adverse weather events. Thus, navigational solutions, such as those proposed in [6,7], rely on the use of automated camera-based systems to improve the navigation and landing of UAVs. However, in confined spaces, such as in warehouse settings, collaborative robots should consider the presence of humans, objects, and manufacturing machines to avoid any potential accidents in the operational space. Therefore, robots need a plan to travel safely to arrive at the final specified target and to avoid any incidents within their navigated path. Navigational environments are the medium in which robots are deployed. Generally, they are different types of robots. Their characteristics vary, such as their size and shape. They possess different capabilities as well, such as the obstacles' avoidance mechanism they use. Given the natural complexity of the environment they operate in, robots require tailored path planning schemes. For example, path planning for indoor environments should consider the existence of walls and narrow channels, such as corridors. Navigating a robot through unknown environments or within indoor manufacturing traffics is prone to several navigational challenges, including those relating to local minima. Such encountered challenges may prevent the robot from reaching its destination or accomplishing its mission. The local minima resulting from common shapes of obstacles, including U-shaped, E-shaped, or V-shaped, constrains the robot from moving forward and reaching its target freely by limiting the navigational area of the robot [8,9]. The problem of local minima emanates when the robot keeps repeating the same steps infinitely. This navigational problem appears mostly in U-shaped obstacles and mazes. Since the robot always follows many steps in the navigation algorithm, it could become stuck in an infinite loop by repeating the steps defined in its algorithm without being able to reach the target destination. This local minima issue is also referred to in the literature as "limit cycle" [10]," deadlock" [11], "dead end", "cyclic dead end", or "trap-situation" [12]. To this end, this paper makes the following contributions:

- An improved and novel algorithm is proposed for the detection and avoidance of local minima.
- The proposed algorithm encompasses five approaches to effectively avoid obstacles, including V-shaped, double U-shaped, C-shaped, and cluttered environments, without falling into the local minima. Mainly, the approaches involve changing the target point temporarily and placing a virtual obstacle covering the local minima region in order to force the robot out of deadlock.
- Several experimental works were set up to evaluate the performance of the five proposed approaches. The results indicate that the Local Path Backtracking approach has the best performance among the five proposed approaches, followed by the Reflected Virtual Target approach.
- Additionally, the results demonstrated that the proposed approaches are quite reliable. For instance, in cluttered environments, the time and distance required to reach a destination by a robot were reduced by eight times when compared to other traditional approaches.
- Overall, the simulation results of the proposed system showed an enhancement in the time required to reach the target in most of the five proposed approaches, especially in the wall-following approach.

This paper is organized as follows: Section 2 describes the main challenges mobile robots face during path planning. Section 3 summarizes the related works. Section 4 describes the base system that the proposed approaches rely on and the fuzzy speed controller employed by the proposed approaches. The proposed approaches to overcome the local minima problem and simulation results are provided in Section 5. Section 6 concludes the paper, and potential future directions are given in this section.

2. Challenges in Online Path Planning

This section is devoted to a brief review of the challenges encountered by robots in path planning. Typically, in online path planning, the robot is expected to overcome several major and minor challenges, such as those reported in [13]. The main challenges are briefly summarized below.

2.1. Obstacle Avoidance

A robot must have a kind of sensory system to avoid collisions with obstacles in the workspace. These systems use various sensors, including ultrasonic sensors, stereo cameras, infrared transceivers, and laser range finding sensors (LIDAR, Light Detection and Ranging). A successful navigation system must be aware of obstacles scattered in the navigation environment at every move; thus, many studies proposed methods to avoid obstacles. The majority of these studies used common approaches in this field, some of which include the use of fuzzy systems [14–16], neural networks [17–19], and Virtual Potential Fields [20–24].

2.2. Goal Seeking, Loops and Speed

Goal seeking is a destination that the robot has to reach at the end. A robot that terminates in a location other than the target is said to have failed its mission. Cyclic dead ends or loops are also one of the serious challenges encountered while designing and implementing a robot navigation system. Controlling the speed of the robot depending on the surrounding environment is equally important as well. The focus of this paper is to evaluate solutions to address these challenges. These possible solutions can reduce the time needed for the robot to reach the target. In real-life applications, such as space robots, robots employed to help rescue missions in catastrophic conditions, and robots deployed in military applications, reducing the time that a robot takes to reach a target point is considered crucial and critical.

3. Literature Review

A recent study discussed the potential and limitations of robots and their connections to other machines to the progress of Industry 4.0 initiatives. Manufacturing, agriculture, kitchen and domestic applications, robotics for healthcare practices, automotive sector, and logistics and warehouse are some of the major potential capabilities of robotics in many industries. Robots are programmed to have a particular amount of intelligence, which is growing as sensor technology improves, not only to do different jobs but also to make decisions, including its adaption in working environments. However, industrial robots need specialized operation and continuous maintenance and development [25]. Whereas Industry 4.0 concentrated mostly on quality, flow, and data collecting, Industry 5.0 focuses more on highly-skilled humans and robots working side-by-side to or even together to develop personalized goods for the consumer. Robots may perform some routine tasks, such as heavy lifting, transportation of raw parts and merchandise, etc., while trained employees focus more on cognitive tasks and creativity. Robots in Industry 5.0 contribute to the reduction in production cost and to an increase in productivity [26]; however, navigational issues, such as path planning and avoiding the local-minima problem in indoor settings, remain amongst the key challenges to their proliferation. On this front, solutions that aim to improve the indoor navigational systems of robots have been previously proposed. The aim is to automate the navigation of Mobile Robots with minimal human intervention. Thus, enabling numerous smart IoT-based applications. For instance, in [27], a multi-sensor fusion approach has been proposed to improve the aerial navigation of robots in industrially-restricted environments. Other works, such as those reported in [28,29] proposed the use of vision-based object recognition solutions to improve the navigation of mobile robots. However, these solutions are generally considered costly and require the use of special sensing devices (e.g., mounted camera and image processing capabilities). Furthermore, ref. [30] proposed a solution to local minima on path planning in unknown environments. In their work, they analyzed patterns in the readings gathered from sensors, where the readings of a sensor compromise two pieces of information; the time when the obstacle is sensed and the location where the robot sensed the obstacle. A similar study in [31] mainly depends on the analyses of spatio-temporal patterns. These patterns were classified to ease recognition of a deadlock situation. The classification also used a two layered-scheme that contains a neural network followed by a fuzzy system. The study shows good performance measured by the length of the path followed by the robot to reach the target point. Another methodology to overcome the local minima problem is proposed in [32]. The local minima situation was defined as a robot following the steps B$\rightarrow$ C $\rightarrow$ B $\rightarrow$ D $\rightarrow$ B, where each of B, C, and D were places already visited by the robot. The solution to the local minima problem was divided into three stages; detection, definition, and avoidance. The environment of navigation was perceived as a grid G, a two-dimensional array. The grid was composed of n square cells, and each cell is represented as C(i)(j). To define the local minima location and size, the proposed approach in [32] built a corresponding map to the grid that showed the explored occupied cells in the grid during navigation. Each cell was represented by a positive integer that incremented each time the robot detected an obstacle occupying the cell. After a local minimum was detected and its enclosure was defined, the robot traveled to a safe destination out of the deadlock enclosure and then closed the enclosure by a virtual wall placed at the entrance of this enclosure, which can be identified by a special laser finding sensor on the robot on the next visit. In addition, some studies [33,34] use the Bug algorithm [35] and its variations to keep the robot away from falling into traps.

Other studies [9,24,36] tried to solve the local minima problem using the wall following approach. This is a popular approach used to get the robot out of a maze by following the walls. This method produced successful results in many environments. However, it suffers from two major weaknesses; firstly, the method fails to reach a target point outside the maze if it follows a wall forming a closed shape. Secondly, the robot must stop following the wall at some point if the target is inside the maze. The success of a method using the wall following technique depends on its ability to determine the point at which the robot should stop following the wall appropriately, for example, in [36], the authors propose a new deadlock detection algorithm that uses the readings of the sensors to determine the size and place of the deadlock. This algorithm keeps running with every set of new sensors' readings to determine the point at which the robot must stop following the wall and be set into target tracing mode again. Although one of its major problems is local minima, many studies use artificial potential field methods for path planning.

4. The Base Navigation System Used in This Work

Previous work proposed an enhanced path planning for mobile robots [9]. They developed a navigation system that uses fuzzy logic and reinforcement learning to emulate a human driver. This section details how the previously proposed path planning system is used as the base navigation system in this work. It provides details on how the five proposed approaches, used to address the local minima and the speed controlling system, were integrated into this base navigation system in the form of separate modules.

The suggested navigation system employs a set of twelve ultrasonic sensors mounted on the robot and carefully aligned to provide greater coverage. Because the front of the robot is the most critical and is the first to encounter impediments, it contains five sensors; the right and left sides each have three sensors, and the backside has only one sensor. Each of these sensors has an "importance" attribute that shows how important the sensor's reading is to the entire robot.

The straight distance between the sensor and the nearest perceived obstruction, ρ, as well as the angle difference between the robot and that barrier, α, are calculated from the sensors' readings. The values of ρ and α are calculated from the readings of three consecutive sensors. These three sensors' values determine the shape of the obstacle. For example, given three consecutive sensors S1, S2, and S3 report three values r1, r2, and r3, respectively. If the values satisfy the inequality r1 < r2 > r3, then the obstacle shape is "channel", as illustrated in Figure 1, [8].

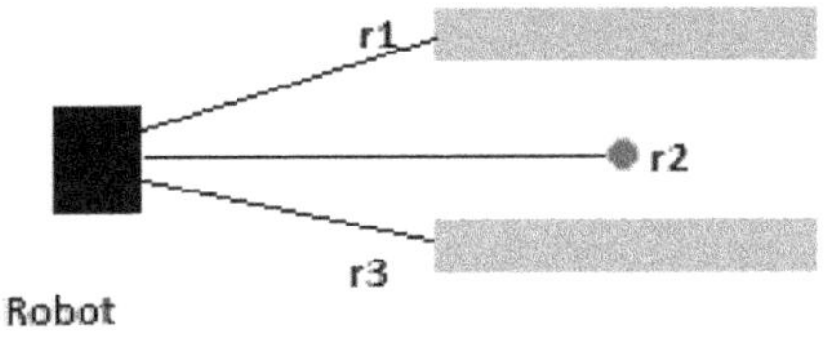

Figure 1. Channel identified by the readings of three sensors.

The fuzzy system uses the values of ρ and α to estimate the value of the angle (ψ_f) at which the robot should travel in the following phase. However, The robot does not perform a straight movement based on this angle. Instead, at time step t, the distance $D_g(t)$ between the robot and the new position indicated by the fuzzy system is measured. The robot then virtually moves to that location and measures the distance $D_g(t+1)$ at the time of step $t+1$, then the difference between $D_g(t)$ and $D_g(t+1)$ is fed into a fuzzy system, which decides the value of $(\Delta\ \psi_f)$. If the proposed angle leads the robot closer to the next target, it is rewarded; if it leads the robot away from the target, it is penalized. After that, the knowledge base, the input of the learner module, is updated. The system considers four actions the robot needs while navigating through the unknown environment; they are as follows:

1. Goal-seeking action: which is responsible for taking the robot to the target point. It includes a fuzzy system that finds the appropriate direction in every step.
2. Obstacle avoidance action: this is responsible for avoiding obstacles. It also depends on a fuzzy system to determine the intensification degree in the difference between the current angle and the angle at which the robot must move to avoid a collision. It depends mainly on how close the robot is to the obstacle. The farther the obstacle is, the smaller will be the angle that the robot has to turn in will be.
3. U-turn action: this action is only activated in two cases: during the initialization phase; if the robot front is not facing the target point, then it must make a U-turn by rotating until it faces the target point. The second case is when the robot gets inside a narrow corridor with a dead end. In this case, it rotates to avoid hitting the walls when it tries to get out of the corridor.
4. Getting rid of local minima action: this action is activated when a local minima situation is detected during navigation. The detection of the local minima situation is performed by finding the average number of U-turns made within a time; if this ratio is high enough to activate this action, the wall following method is called to take the robot out of the trap. The robot follows the nearest wall it detects and keeps walking close to the wall while overlooking the attraction force of the target point for some time. After that time, the robot returns to the goal-seeking action and disables the wall-following action. If the robot finds that it is again trapped in the same local minima, then the time of wall-following is extended.

As we mentioned before, not only the problem of local minima must be addressed in path planning systems, but also the robot must navigate with controlled speed; that is to balance the cost of the robot tasks and the safety of the working environment and the robot itself. In [8], the problem of speed control is discussed and addressed using a fuzzy speed control system. The study suggests a fuzzy controller that controls the speed of the robot depending on three factors; the turning (heading) angle (θ) in every step taken by the robot, and ρ in (Equation (1)), which is the distance to the nearest obstacle sensed by the ith sensor (imp_i) in (Equation (1)), and the importance of the ith sensor's reading (D_i) in the same equation. In every step, each of (θ) and ρ is found, and the robot adjusts the speed accordingly. ρ is calculated using the following formula:

$$\rho = \frac{imp_i}{D_i} \quad (1)$$

The importance of the sensor's reading describes how important this reading is according to the position of the sensor on the robot. For example, the frontier sensors are more important than the rear sensors as the robot only moves forward. The importance of each sensor is represented with a value in the range (0–1). After finding the values of θ and the highest ρ among all of the sensors, they are entered into a fuzzy inference system to find the final value of the speed for the next move. The final value of the speed is measured by the length of the step in the next time-step $(t + 1)$.

The fuzzy Inference System (FIS) is built on 25 rules to control the robot speed. The FIS takes two inputs; θ which is the angle between the robot and the near obstacle, and ρ which is the distance between the robot and the obstacle. Every one of the inputs has five possible values (i.e., θ can be Very Small (VS), Small (S), Moderate (M), Large (L), or Very Large (VL), and ρ can be Very Low (VL), Low (L), Moderate (M), High (H), and Very High (VH)). The FIS output is the appropriate speed of the robot in meters/second, which varies between 0 and 1. For more details on rules and membership functions, the reader is referred to [8].

5. Addressing the Local Minima Problem by Target Switching

This section introduces the novel algorithm proposed in this work. The algorithm encompasses approaches that aim to detect the local minima and approaches to get the robot out of the deadlock enclosure. All of the approaches use the same proposed detection algorithm because of its precision and ability to detect the trap situation effectively. Moreover, the five approaches with the detection algorithm were compared to the wall-following approach used in [9]. The same base navigation system proposed in [9] was used for obstacle avoidance and goal-seeking.

5.1. Environment Perception

The navigation environment was perceived as a grid composed of n cells. One major problem that should be discussed is the partially occupied cells. This problem arises in most navigation systems that depend mainly on grids when the robot cannot move to a partially occupied cell, even when there is not enough space for the robot to traverse this cell. The problem is illustrated in Figure 2. To avoid this problem, the robot does not depend on the grid cells to find the next move. Instead, the robot uses an independent navigation system proposed in [9] to navigate between obstacles. This means that the grid is only used for local minima detection and avoidance while using the independent navigation system, the robot keeps tracking and updating location information in terms of cells. In [9], the action taken by the robot at any time depends on the immediate mapping of the ultrasonic sensory data. However, the robot has no memory or fuzzy speed mechanism where our contribution addresses these important features.

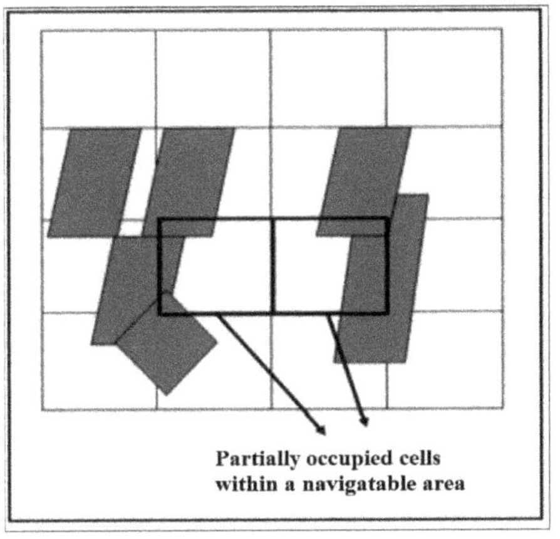

Figure 2. The problem of partially occupied cells.

The number of cells (n) in the grid can be found using Equation (2):

$$n = \left\lceil \frac{A}{S_R} \right\rceil \quad (2)$$

where n is the number of the cells in the grid, A is the rectangular area of the environment, and S_R is the size of the robot. This means that there are $\frac{L}{R_L}$ cells, where L is the length of the environment, and R_L is the length of the robot. The robot in [9] gets its location at every movement step using the turning angle, and the distance from the starting point. This location information can be used to find the robot's location in terms of a cell using the following formulas:

$$C_x = \left\lceil \frac{x}{R_L} \right\rceil \quad (3)$$

$$C_y = \left\lceil \frac{y}{R_L} \right\rceil \quad (4)$$

where in C_x (Equation (3)) and C_y (Equation (4)), x and y are the x-coordinates and y-coordinates, respectively, for the robot's current location. At each location update, the robot found its current location within the grid and stored the cell's index in the *Vistited_Cells* vector. For example, the visited cells in Figure 3 are given by a vector of spatial data (2,5), (3, 4), (3,6), (3,7), (4,7). Note that the icon of the robot in Figure 3 becomes a dark square when the local minima algorithm is activated, and a circle in an independent navigation system, such as in [9]. When traversing a cell (i), the robot tests three situations:

1. If the cell does not exist in Visited_Cells, then it is added to the vector, and the number of visits for the current cell NV(i) is increased by 1.
2. If the cell already exists in the vector, and the robot is still traversing the same cell with multiple steps (i.e., within the same cell's borders), then do nothing.
3. If the cell already exists in the vector, and the robot traverses it for the ith time, then the number of visits for the cell is incremented by 1.

In this way, the problem of early and erroneous detection in [32] is solved. The grid perception of the environment was used for the following reasons:

1. Initially, the robot does not memorize the occupied cells. The robot cannot be precise in checking whether the place is visited or not depending on point perception (x,y) of the environment, as shown in Figure 2.
2. If the robot detects deadlock situations, it needs to remember how many times it visits a region to detect the local minima.

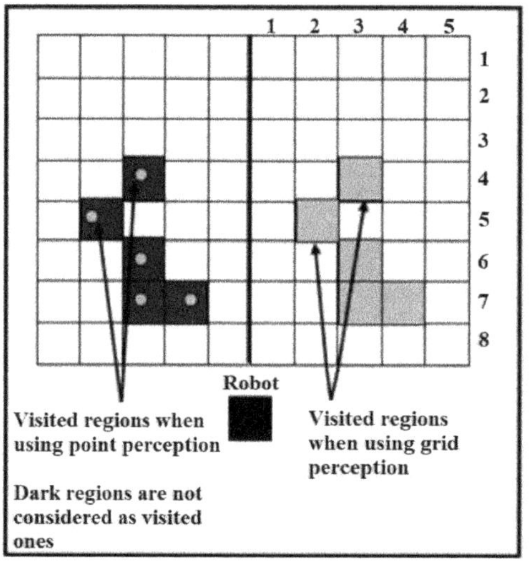

Figure 3. The difference between the effectiveness in cell perception and point perception.

5.2. Local Minima Detection

The robot used the grid to detect the local minima situation. In every step, the robot finds the value of two variables; the local minima, or Deadlock chance (*D*), and the Threshold (*T*), using the following equations:

$$D = Adjacency \times Intensity \tag{5}$$

$$T = \frac{R_T}{L_D} \tag{6}$$

$$Adjacency = \frac{G}{c_D} \tag{7}$$

where *G* is adjacent to the revisited cells and c_D is the revisited cells.

$$Intensity = T \times \frac{R}{V_0} \tag{8}$$

where R_T is the total number of revisits events in the grid, *R* is the total number of cells with multi-visits, and V_0 is the total number of cells with only one visit within the grid. The *Adjacency* (Equation (7)) is the factor used to express how close the revisited cells are to each other, and the *Intensity* (Equation (8)) measure is used to describe how intense the revisiting is in the current region. Both *Adjacency* and *Intensity* are found using Equations (7) and (8), respectively. In Equation (6), L_D refers to the length of the deadlock, which is the length of the rectangular area that encompasses all the revisited cells. Note that the cells with a number of visits greater than 1 and not adjacent to other revisited cells (i.e., clusters composed of a single cell), are all neglected. As noticed from the above formulas, the *Adjacency* is affected by the distance between the clusters and the number of these clusters. When the distance between clusters increases, their *Adjacency* decreases. In addition to that, *Intensity* increases each time the robot traverses the same groups of cells and when more cells are revisited. The algorithm of local minima detection uses the previous formulas to determine whether the robot must activate the mode "get out of trap" or not. Once the deadlock is detected, the algorithm tries to explore and define the deadlock enclosure. The process of evaluating whether a robot is in a dead-end situation or not is called a deadlock detection algorithm. Dead-end situations happen when there is no free obvious path between the robot and its target. The method *Define_Deadlock* is invoked to define the obstacle(s) that form the deadlock enclosure. While navigating, the robot keeps track of the occupied cells by storing the occupied cell's index in the *Occupied_Cells* vector. The method *Define_Deadlock* uses this vector to determine which cells form the deadlock enclosure. The method first finds the nearest occupied cell in the vector *Occupied_Cells* to the robot. Then, starting from that cell, it explores the adjacent cells that are occupied too, until reaching an End Cell. An End Cell is a cell that is occupied and adjacent to only one occupied cell, as shown in Figure 4. The two algorithms below demonstrate the process of identifying and detecting the deadlock, which is the first step in determining the presence of the local minima problem. Following the detection and identification of the problem, a variety of strategies are explored in order to solve the problem, which will be presented in this study. These algorithms were given special attention because they represent the foundation of the problem and the solutions that are being assessed in this study.

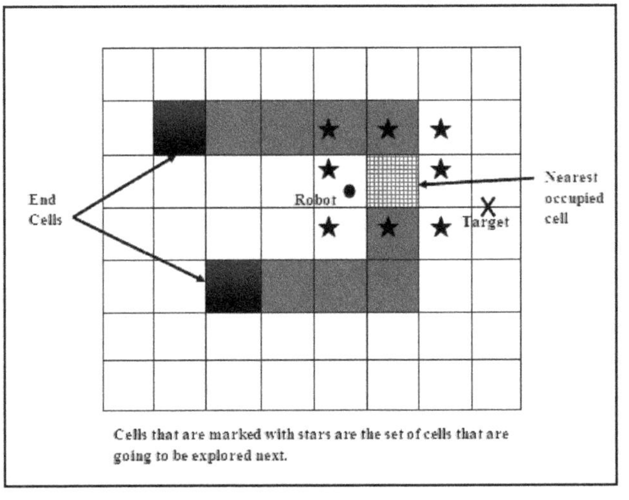

Figure 4. *Define_Deadlock* Method.

In each step of exploration, *Define_Deadlock* takes one cell as the center cell and then explores the eight adjacent cells. If any cell in the adjacent eight is occupied, it is stored in the *Deadlock_Enclosure* vector and the Exploration Stack. In the next step of exploration, the top cell in the Exploration Stack is popped and set as the center cell, then explored. The method keeps repeating these steps until it reaches the end of the enclosure. In this way, it can define the deadlock enclosure precisely and determine the cells that form it. The algorithm of the local minima detection and definition works as follows: considering the number of navigation steps S, the algorithm of local minima detection and definition has a linear time complexity of O(S) in its worst case, and that is when the robot detects/defines a local minima in every step and when all of the discovered occupied cells are pushed into the stack. After defining the deadlock enclosure, the robot will easily get out of it using the end cells. An end cell is found by counting its surroundings from the vector deadlock enclosure; if the number of surroundings equals 1, then it is an end cell.

```
Deadlock Detection {
Find Intensity, Adjacency, Threshold, Deadlock Chance
If (Deadlock Chance > Threshold):
  Call Define Deadlock
  Activate "get out of trap"
}

Define Deadlock
{ Find the nearest occupied cell N to the robot
Center <- N
Stack.Push(Center)
Stack.length <- 1
Repeat until stack.length=0
  find adjacent cells coordinates
  Adjacent[0]=(Center.x−1,Center.y) // left
  Adjacent[1]=(Center.x+1,Center.y) // right
  Adjacent[2]=(Center.x−1,Center.y−1) //upper left
  Adjacent[3]=(Center.x−1,Center.y+1) //lower left
  Adjacent[4]=(Center.x,Center.y−1) //upper
  Adjacent[5]=(Center.x,Center.y+1) //lower
  Adjacent[6]=(Center.x+1,Center.y−1) //upper right
  Adjacent[7]=(Center.x+1,Center.y+1) //lower right
  For i = 0 To 7: // adjacent cells indexed from 0 to 7
    if Adjacent[i] is in Occupied
    Stack.PUSH (cell i)
    Stack.Length
    Stack.Length+1
    Insert cell i into Deadlock-Enclosure
  Next i
Center <- stack.POP
Stack.Length <- Stack.Length-1
Loop }
```

5.3. *Addressing the Local Minima*

Once the deadlock enclosure is detected and defined, the mode "get out of trap" is activated. In this paper, there are five different approaches proposed to get out of the trap. All of these approaches use the same detection algorithm described in the previous section. Mainly, these approaches depend on two things: 1. Placing a virtual target in appropriate locations instead of the real one to address the target attraction force effect on the robot. 2. Placing virtual obstacles on the deadlock enclosure. This prevents the robot from falling into the same trap after getting out of it.

5.3.1. Random Virtual Target

In this approach, a virtual target point is placed in a random location within a limited area around the farthest end cell from the target. The robot is affected by the new target attraction force and stops seeking the real goal. As a result, the robot starts heading to the virtual goal until it reaches it. Once the robot reaches the virtual goal, the real target point is set back as the target. To avoid the robot being trapped in the same deadlock enclosure again, a virtual obstacle is placed over the whole deadlock enclosure. Closing the deadlock enclosure is performed by calling the *Close_Deadlock* method. This method first checks whether the robot is still within the area that is to be closed. If so, the virtual obstacle keeps shrinking in a constant ratio until the robot is out of the closed area. The virtual obstacle dimensions are within the coordinates X_S, Y_S, X_L, and Y_L, which are the smallest x-coordinate, the smallest y-coordinate, the largest x-coordinate, and the largest

y-coordinates, respectively, among the coordinates in the *Deadlock_Enclosure* vector. The random virtual target selection approach is illustrated in Figure 5.

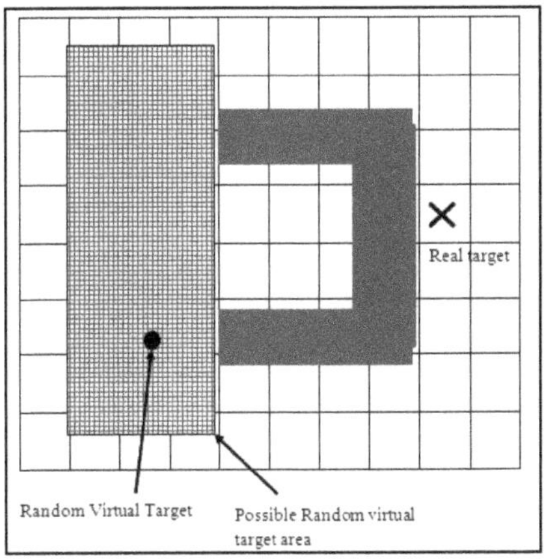

Figure 5. Random virtual target selection.

To avoid placing the virtual target in a location occupied by an obstacle, the robot first checks whether the selected location for the virtual target is located on an occupied cell using the vector *Occupied_Cells*. If it is found that an obstacle occupies the selected location, it chooses another random virtual target location.

5.3.2. Reflected Virtual Target

This approach mainly depends on the fact that in most local minima situations, the deadlock enclosure's opening faces the robot on its way to the target located behind that enclosure. In this case, the problem can be solved by placing a virtual target in front of the obstacle and within the area that precedes the enclosure opening. Once the robot reaches the virtual target, it switches back to the real target and calls the method *Close_Deadlock*. This can be achieved by simply making a reflection of the real target near the enclosure's exit. The real target is reflected either horizontally or vertically depending on the enclosure's exit direction. The appropriate reflection of the real target is achieved by assuming the middle of the enclosure as the reflection axis. This approach is illustrated in Figure 6.

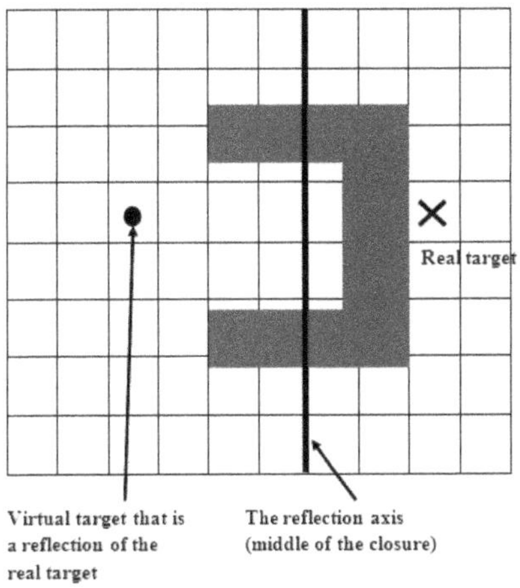

Figure 6. Virtual target by reflection of the real target.

5.3.3. Backtracking

One way that guarantees the escape from the deadlock enclosure is that the robot follows its steps back to the exit, which is the entrance to the enclosure. Since the visited cells are all stored in the *Visited_Cells* vector, the robot can easily backtrack its path. To disable the real target's attraction force, a set of virtual targets are placed on the visited cells that form the backtracking path, starting from the robot's current location inside the deadlock enclosure and moving backward until a "stop backtracking" point is reached. The sequence of virtual targets is a set of virtual targets placed at every constant number of cells in the *Visited_Cells* vector. The "stop backtracking" point can be determined using one of three following:

1. **Global Path Backtracking:** The stop point is the same as the start point (S) of the navigation. The robot keeps backtracking until it reaches the starting point. This approach is effective in the case of small environments. However, it is inefficient in wide environments because the robot must reach the very distant starting point when it encounters a deadlock. After the robot escapes from the deadlock, and the mode "get out of trap" is disabled, the start point is re-initialized and set as the first cell the robot traverses after closing that deadlock enclosure.
2. **Half Path Backtracking:** The stop point is the midway point between the current locations of the robot and the starting point. This approach is more effective than the previous one in wide environments but less effective in small environments because the stop point could be inside the enclosure of the deadlock.
3. **Local Path Backtracking:** The stop point is at the end cells. This one could be the most appropriate choice for the point of "stop backtracking" as it guarantees that the robot will not travel so far. On the other hand, it guarantees the robot is out of the deadlock enclosure. The three approaches for choosing the "stop backtracking" point are shown in Figure 7. After the robot reaches the last virtual target in the sequence, a square virtual path of straight lines is set all around the deadlock enclosure. Then, a final virtual target point is set on this path. The virtual path must pass by the real target point and near the last virtual target from the backtracking sequence, as illustrated in Figure 8. The final virtual target is determined as the middle point of the distance between the robot and the real target. In the case that there is not enough space for the virtual path (i.e., not enough space under or above the enclosure for the robot to move), then the final virtual goal is placed on the opposite side of the square virtual path. Final virtual target placed on the virtual path, guaranteeing that the robot moves towards the real target and away from the deadlock.

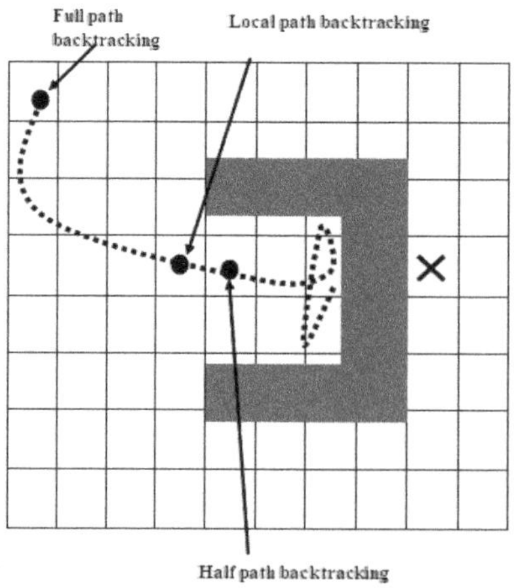

Figure 7. Three approaches to determine the backtracking stop point.

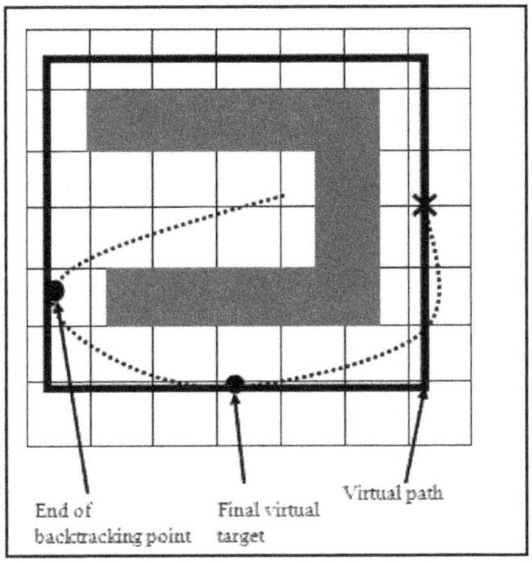

Figure 8. The virtual path in the backtracking method.

5.4. Simulation Results

This section is devoted to reporting on the simulation works conducted in this work. Each of the five proposed approaches was evaluated, and their performance was investigated under different setups. There are four different test cases conducted in this section, including path planning in the presence of C-shaped obstacles, double U-shaped, V-shaped obstacles, and when the robot encounters cluttered environments on its path. The results of each test of the five approaches proposed to address the local minima are then discussed and analyzed.

The unit used to measure the efficiency of the proposed methods to address the local minima is the number of steps the robot makes while traveling from the start point to the target point. The size of a step is constant and equals 10 cm. There are general parameters used in the simulation of the proposed approaches to address the local minima and the fuzzy speed controller. The parameters are summarized in Table 1.

Table 1. Simulation Parameters.

Constant-Speed Robot	Speed: 0.5 m/s	**Fuzzy-Speed Robot**	Speed: 0–1 m/s
	Step length: 0.1 m		Step length: 0.0375–0.2125 m
	Step time: 200 ms		Step time: 200 ms
	Sensing range : 4 m		Sensing range: 4 m
Simulation Environment	Robot size: 0.7 × 0.7 m		
	Environment area: 14 × 24 m		
	Operating System: Microsoft Windows		

Local Minima Avoidance

Figure 9 depicts the performance of detecting and identifying the enclosure of a deadlock. The little circles on the cells show whether the cell is occupied or near to another cell that is occupied. This is dependent on the robot's sensing range; as seen in Figure 9, the above barrier is recognized near the robot, while the below horizontal light green region has yet to be discovered by the robot's sensors.

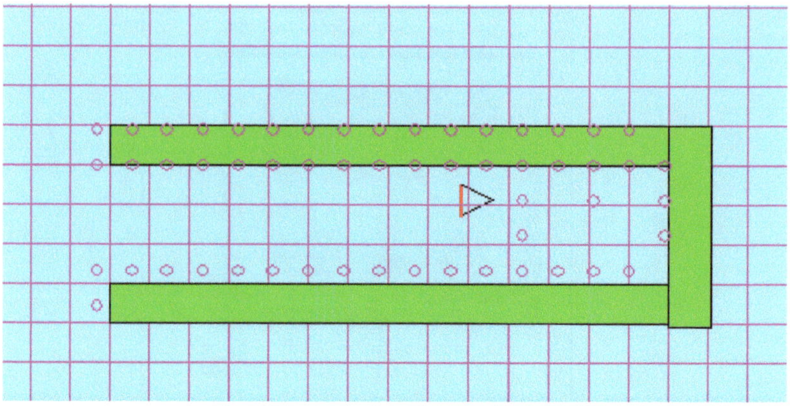

Figure 9. Deadlock detection and definition performance.

There are four different test cases conducted on the five proposed approaches in addition to the wall-following approach for comparison.

Test case #1: C-shaped obstacle test case

In this test case, the robot is in front of a C-shaped obstacle. Figure 10a–f shows the approach's performance. The black area represents the visited path by the robot. The challenge in this test case is that the C-shaped obstacle is a circle with a small exit. The performance of the wall-following approach, in Figure 10f, is the worst compared to other proposed approaches. It also required following the whole environment's perimeter, which is inefficient in large environments. Figure 10a,b shows the superior performance of the reflected target over the random virtual target. This is due to the location of the chosen virtual target. In the reflected virtual target, the virtual target is a reflection of the real target position vertically. The reflected virtual target is near the lower end of the C-shaped obstacle, unlike the random virtual target chosen near the upper end of the obstacle. This added a few extra steps in the random virtual target approach because the only way to reach the target is under the C-shaped obstacle. Figure 10c,e shows the robot's similar performance, which is because of the similar location of the "stop backtracking" point. In Figure 10d, we noticed the less efficient performance of the robot when it stops backtracking in the middle point of the path; this is because the middle point, in this case, is located inside the C-shaped obstacle. Thus, it required more than one virtual obstacle to close the deadlock enclosure and leave it.

Test case #2: Double U-shaped test case

In this test case, the robot enters a trap that is nested in another trap. Figure 11a–f shows the performance of the six approaches. Figure 11a,b shows the advance in performance for the random virtual target approach over the reflected virtual target approach. This can be explained by the way the virtual target is chosen. In the reflected virtual target, the robot finds itself trapped in the inner deadlock, so it places a virtual target that is a horizontal reflection of the real target. This made the virtual target location to be near the inner's deadlock left end. When the robot switched back to the real target, it was still in the same location as the virtual target; the robot entered the outer deadlock again from the left, then detected that it was trapped again after reaching the right side of the deadlock's enclosure. After that, it switched back to the real target again, which is the same reflection, but over the middle line of the outer deadlock this time. When the robot switched back to the real target, it was still on the right side while the virtual target was in front of the deadlock but on the left side, which required the robot to go back to the left side again, then go out of the enclosure to the reflected target which adds a few more steps. Unlike this approach, the random choice of the virtual target came once on the left side and once on the right side of the enclosure. As noticed in Figure 11c, the robot stops backtracking when it reaches the starting point. The starting point for the second outer deadlock is the same point where the robot was upon closing the first inner deadlock. Thus, the robot had to reach that point first before closing the outer deadlock, which added a few extra steps for the robot to reach the real target compared to the approaches in Figure 11d,e. In the half path backtracking approach of Figure 11d, the robot stops backtracking when it reaches the middle point of the path, and in Figure 11e, the robot stops backtracking when it reaches the enclosure exit. This is the reason for the similar performance of the local path backtracking and half path backtracking. In Figure 11f, we notice again the inefficient performance of the wall-following method, which caused the robot to traverse the target's region without seeing it due to the disabling of the goal-seeking action while following the wall. It is important to clarify that the magenta color has been used just to emphasize that the robot does not follow a specific color in Figure 11, and that is also applicable to other figures; Figures 10, 12, and 13.

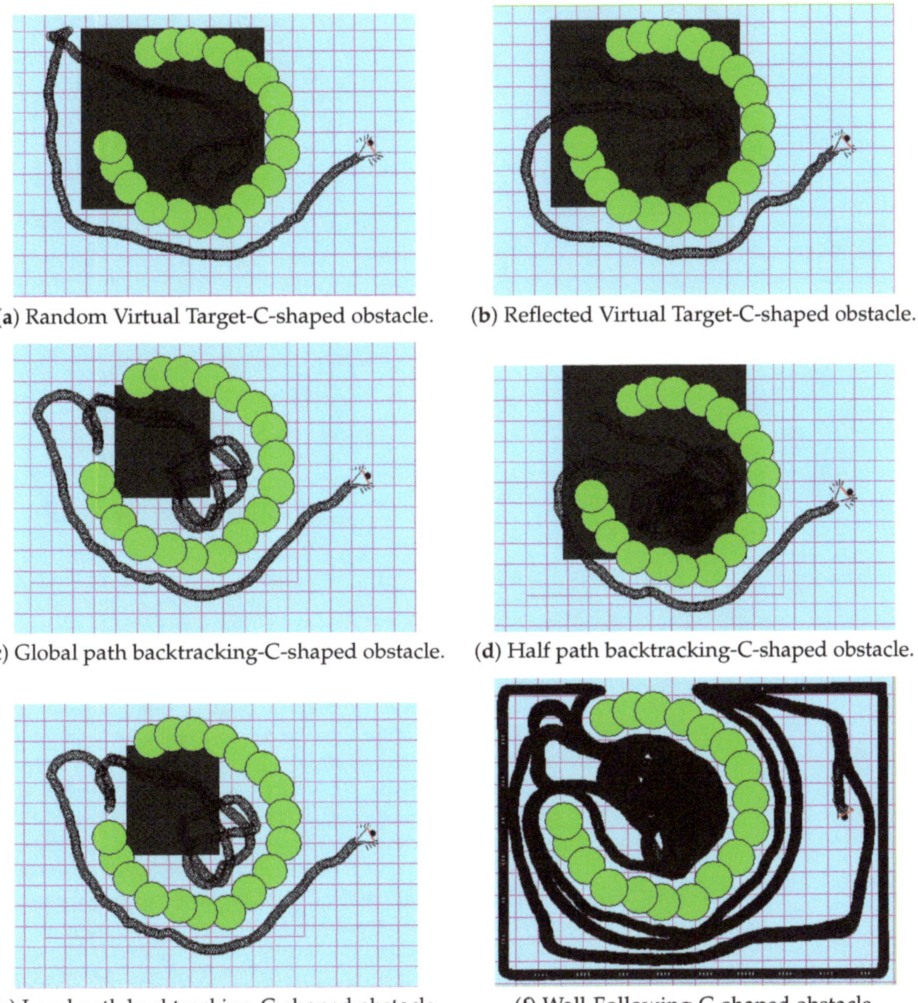

(**a**) Random Virtual Target-C-shaped obstacle. (**b**) Reflected Virtual Target-C-shaped obstacle. (**c**) Global path backtracking-C-shaped obstacle. (**d**) Half path backtracking-C-shaped obstacle. (**e**) Local path backtracking-C-shaped obstacle. (**f**) Wall-Following-C-shaped obstacle.

Figure 10. C-shaped obstacle test case.

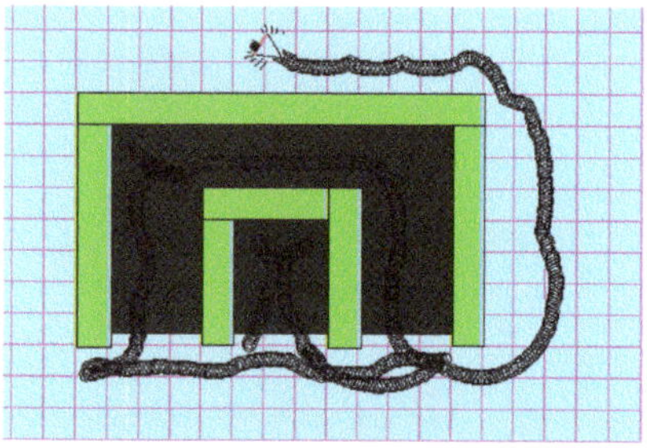

(**a**) Random Virtual Target-Double U-shaped obstacle.

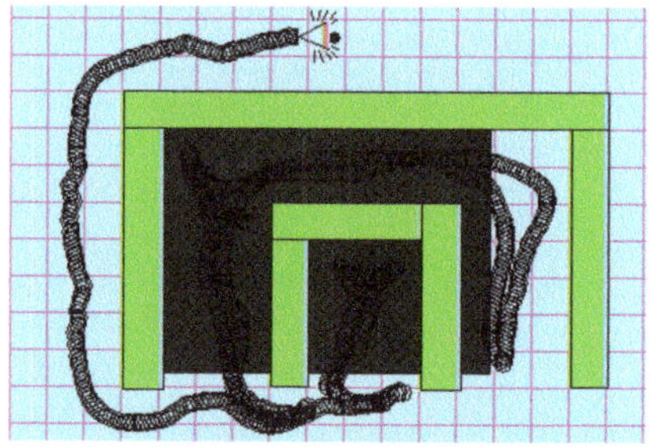

(**b**) Reflected Virtual Target-Double U-shaped.

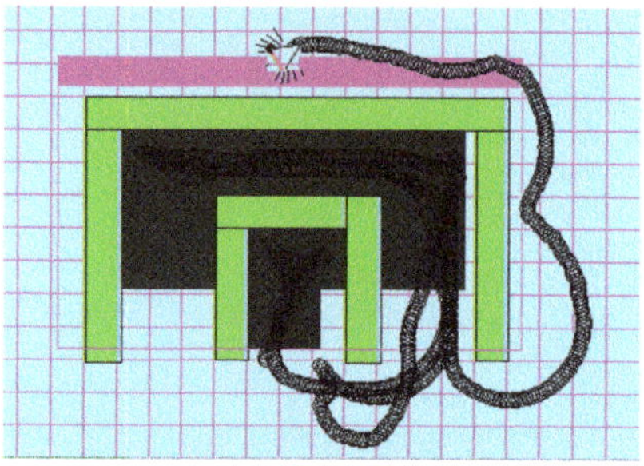

(**c**) Global path backtracking-Double U-shaped obstacle.

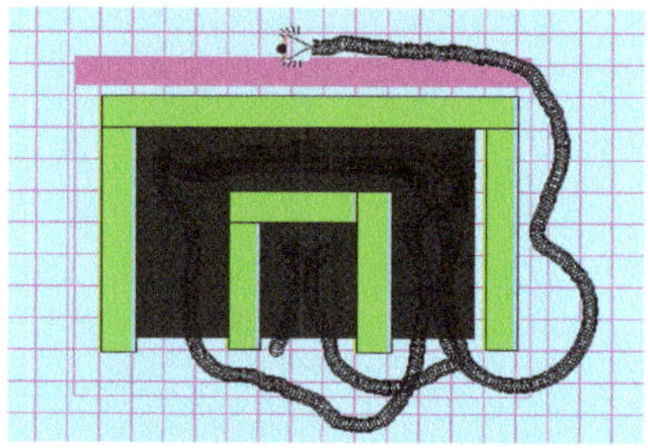

(**d**) Half path backtracking-Double U-shaped obstacle.

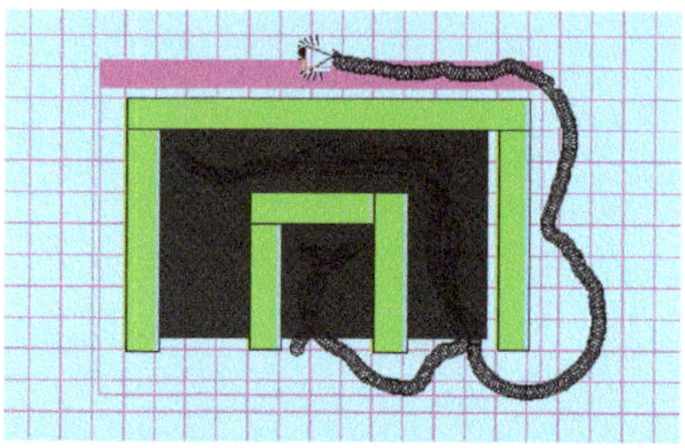

(**e**) Local path backtracking-Double U-shaped obstacle.

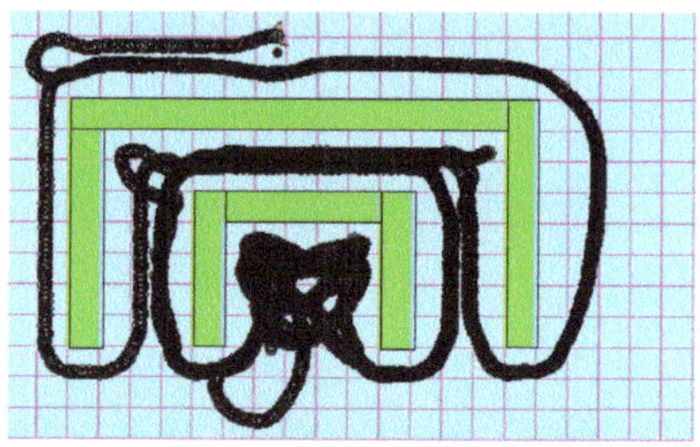

(**f**) Wall-Following-Double U-shaped obstacle.

Figure 11. Double U-shaped test case.

Test case #3: V-shaped test case

In this test case, the robot enters a V-shaped obstacle. Figure 12a–f shows the performance of the six approaches. In Figure 12a, the robot closed most of the enclosure in the first round, but not all of it because the random target at the first round was placed near the end of the enclosure but almost inside. Thus, the robot fell into the same deadlock twice and required two virtual obstacles to close the deadlock region. However, in Figure 12b, we notice a better performance of the reflected target approach because the reflection of the real target is outside the enclosure, which required one round to close the whole deadlock. In Figure 12c,e, we notice a similarity in the performance of the global path and local path backtracking. Because the "stop backtracking" points are almost in the same location (i.e., the robot started navigation from a point near the exit of the deadlock), the performance was different in the half path backtracking approach of Figure 12d. In this test case, this point was inside the V-shaped obstacle; thus, the virtual obstacle shrunk to avoid closing on the robot inside the deadlock. In Figure 12a–e vs. Figure 12f, the wall-following approach took the most significant number of steps, the longest path, to get out of the trap situation because it forced the robot to follow part of the environment's perimeter wall.

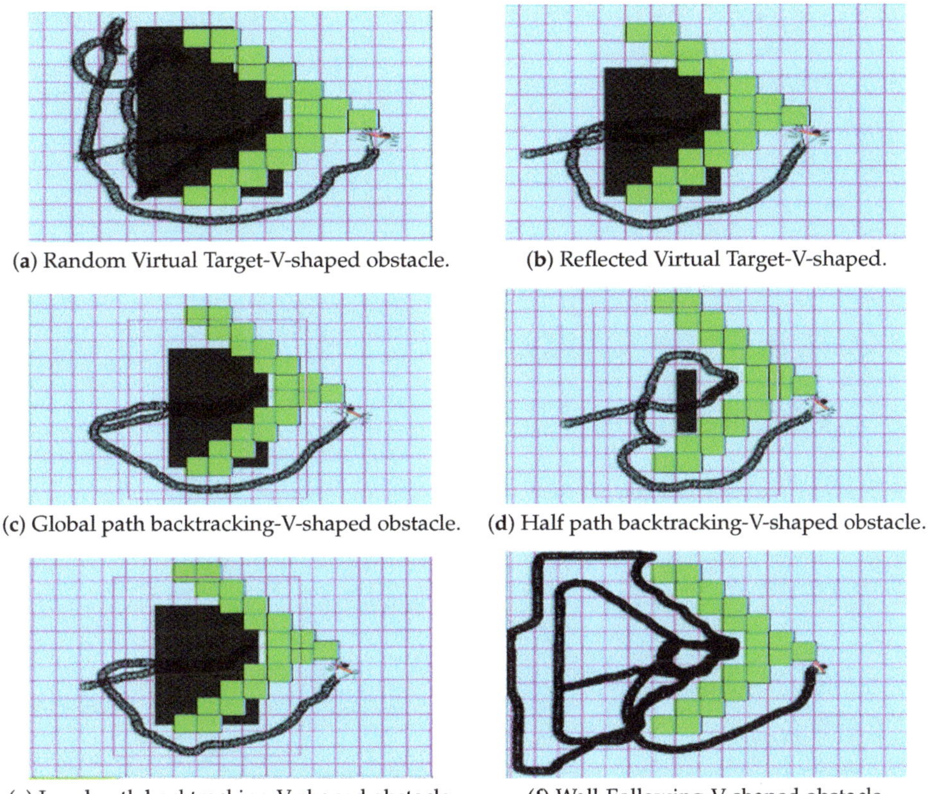

(a) Random Virtual Target-V-shaped obstacle.

(b) Reflected Virtual Target-V-shaped.

(c) Global path backtracking-V-shaped obstacle.

(d) Half path backtracking-V-shaped obstacle.

(e) Local path backtracking-V-shaped obstacle.

(f) Wall-Following-V-shaped obstacle.

Figure 12. V-shaped test case.

Test case #4: Cluttered environment test case

This is the last test case, and it shows the performance of the six approaches in cluttered and crowded environments. Figure 13a–f shows the results of the test case. In Figure 13a,b, the random target choices, either the random virtual or the reflected virtual targets, show the best performance compared to others. In Figure 13c, we notice the behavior of global path backtracking. The starting point is located a bit far from the deadlock. When the robot needs to get out of the deadlock, it must return to that far point, which is inefficient in such cases. This problem is alleviated in the half path approach, as illustrated in Figure 13d. However, it still adds extra unnecessary steps because the deadlock was not fully covered with the virtual obstacle. This happens when the virtual target is located inside the area that must be covered. So, the virtual obstacle keeps shrinking until the robot is uncovered. The best of the "stop backtracking" points is near the exit of the deadlock enclosure achieved by the local path backtracking approach, and this is clear in Figure 13e. In this test case, Figure 13f shows that the wall-following approach performance performs inefficiently to get to the target. The robot was forced under this approach to circumnavigate around some obstacles (closed shapes) many times, thus wasting a lot of time. Table 2 shows the path length in meters that was taken by the robot working with the six approaches to reach the target point. As noted in the table, the robot with the wall-following approach has the longest path in all of the four tests.

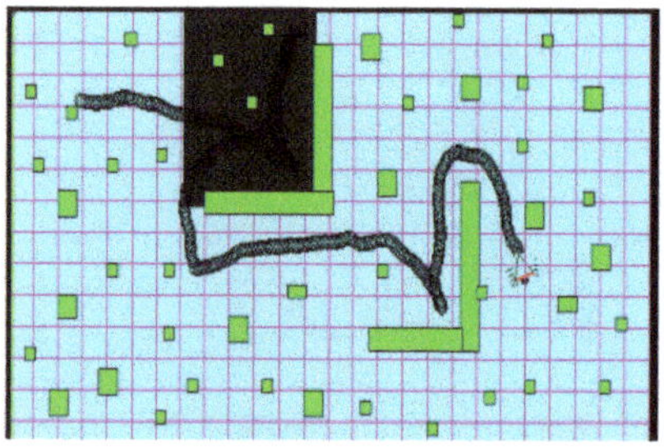

(**a**) Random Virtual Target-Cluttered environment obstacle.

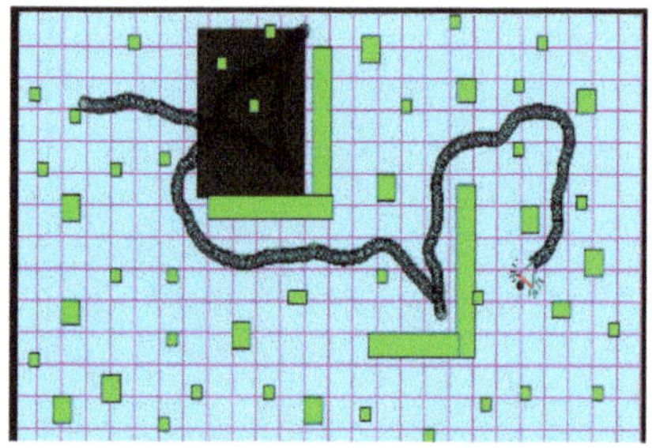

(**b**) Reflected Virtual Target-Cluttered environment.

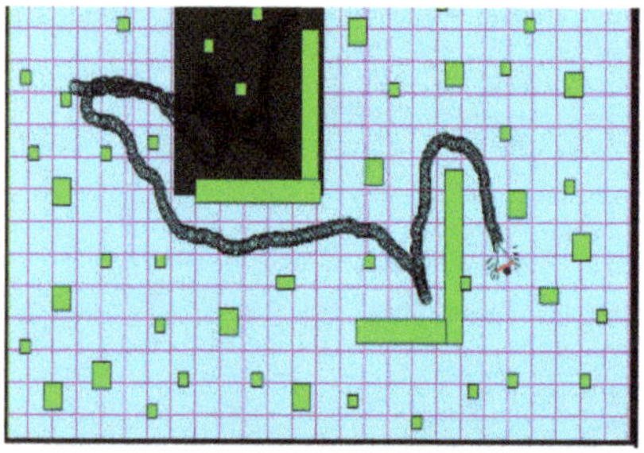

(**c**) Global path backtracking-Cluttered environment.

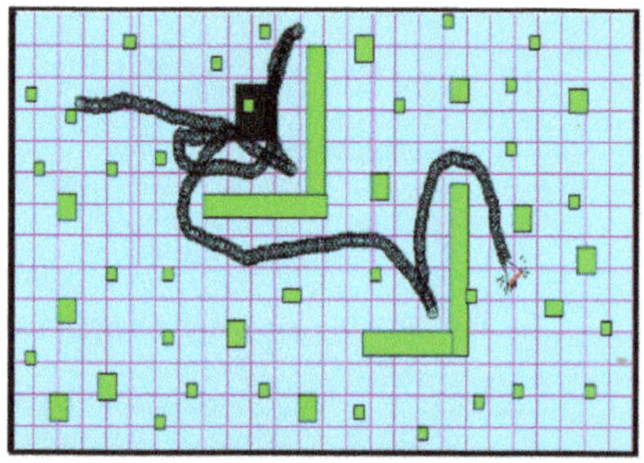

(**d**) Half path backtracking-Cluttered environment.

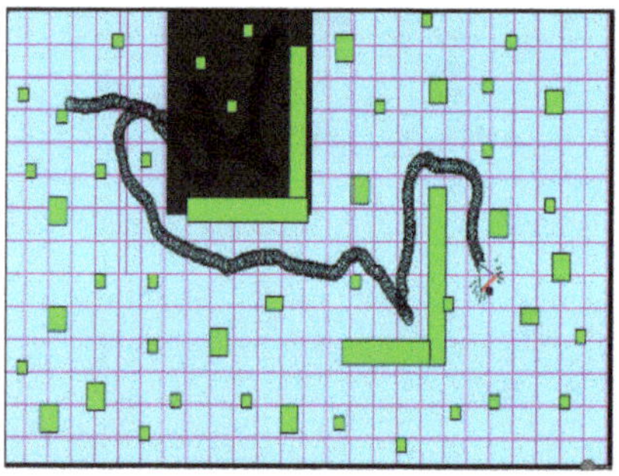

(**e**) Local path backtracking-Cluttered environment.

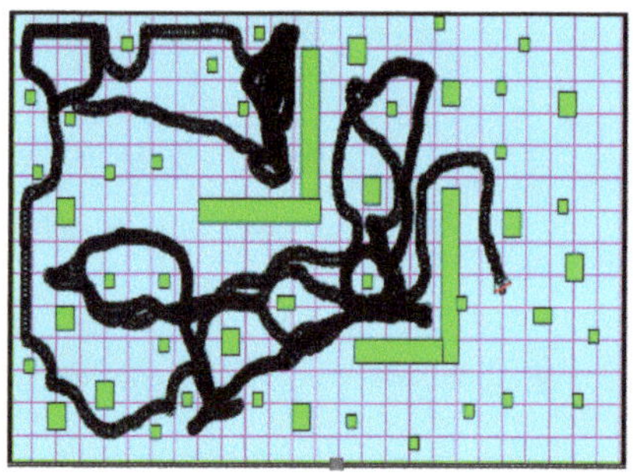

(**f**) Wall-Following-Cluttered environment.

Figure 13. Cluttered environment test case.

Table 2. The distance of robot routes for Figures 10–13 (measured in meters).

Approach Testcase	Random	Reflected Virtual Target	Global Path Backtracking	Half Path Backtracking	Local Path Backtracking	Wall-Following
C-shaped	55	45	59	101	59	1915
Double U-shaped	97	88	100	110	96	466
V-shaped	38	27	29	31	28	111
Cluttered	46	47	55	47	52	436
Average	59	66.25	60.75	72.25	58.75	732

The reflected virtual target shows better performance in test cases 1 and 2, and this is due to its optimal choice for the target location in these test cases. As noted in the table, the test case that required the longest path in all six approaches is the test case of the double U-shaped obstacle.

The chart in Figure 14 shows the performance of the five approaches compared to the wall-following approach. The efficiency of the six approaches is measured in the number of steps required for the robot to reach the final real target. One important point to be mentioned in this section is that each experiment was conducted 10 times to mitigate the effect of randomization. The results recorded in this section are an average of ten runs.

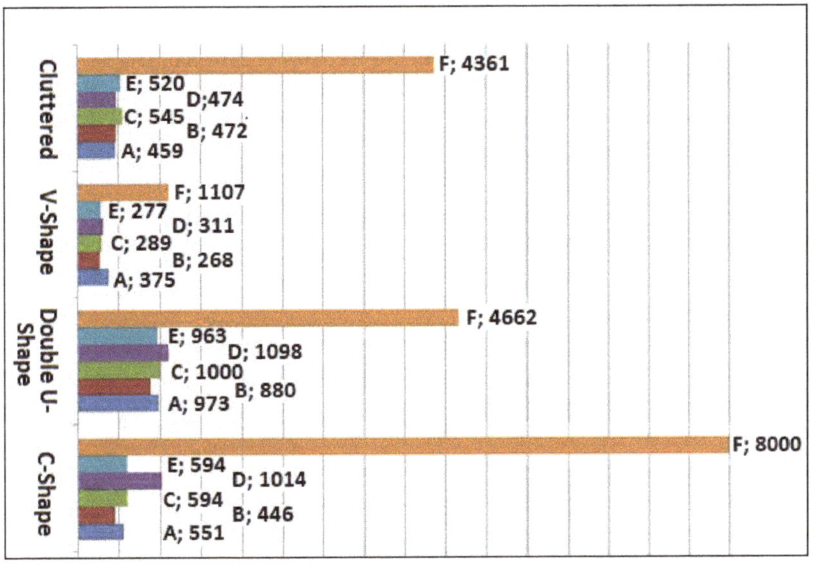

Figure 14. Performance of the proposed approaches to overcome the local minima compared to the wall-following approach (measured by the number of steps).

5.5. Speed Control Effect on the Five Proposed Approaches to Address the Local Minima

This section reports on the studies conducted to determine the effect of controlling the speed using the previously proposed fuzzy controller [8] on the performance of the proposed approaches. To address the local minima problem, the mobile robot of fuzzy speed was provided with these approaches and then tested using the same four environmental setups by replicating the same encountered local minima. Table 3 shows the efficiency performance of the five proposed approaches to address the local minima with constant speed within the four environments. Table 4 shows the performance of the five proposed approaches with a fuzzy-speed robot. The performance of the proposed approaches is measured by the time (in seconds) elapsed between the start and the end of navigation. It shows the wall-following approach consumes the longest time in the trips. The local path backtracking approach seems to be the best choice for leaving the deadlock enclosure in general due to its reasonability in choosing the right point to stop backtracking. On the other hand, half path backtracking seems to be the least suitable approach when the robot starts navigating from a location that is close to the deadlock enclosure, unlike global path planning, which is suitable for short paths but not for long ones.

Table 3. Time spent during navigation by the five proposed approaches to address the local minima with a constant-speed robot (seconds).

Approach Testcase	Random	Reflected Virtual Target	Global Path Backtracking	Half Path Backtracking	Local Path Backtracking	Wall-Following
C-shaped	110.2	89.2	118.8	202.8	118.8	1600
Double U-shaped	194.6	176	200	219.6	192.6	932.4
V-shaped	75	53.6	57.8	62.2	55.4	221.4
Cluttered	91.8	94.8	109	94.8	104	872.2
Average	117.9	103.4	121.4	144.85	117.7	906.5

For the U-shaped, W-shaped, and E-shaped obstacles, the best approach that can be used to take the robot out of the deadlock enclosure is the local path backtracking. For the double U-shaped and V-shaped obstacles, the best choice is to use the reflected virtual target approach, especially if the target is located right behind the deadlock enclosure and the robot in front of it. The random virtual target approach is the most suitable approach for complex environments, such as cluttered environments.

Table 4. Time spent during navigation by the five proposed approaches to address the local minima with a fuzzy-speed robot [2] (measured in seconds).

Approach Testcase	Random	Reflected Virtual Target	Global Path Backtracking	Half Path Backtracking	Local Path Backtracking	Wall-Following
C-shaped	96.8	71.8	98.6	223.2	97.8	80
Double U-shaped	75.6	149.6	104.4	223.6	252.2	169.8
V-shaped	83.2	48	54.6	59.6	51.2	400.4
Cluttered	127.6	105	132	88.4.	132.	1378.8
Average	95.8	93.6	97.4	168.8	133.3	507.25

When the robot moves under the proposed fuzzy speed control, the time needed to reach the target is decreased in general, although there are some odd cases, such as the majority of the approaches' performance in the E-shaped obstacle test and in the double U-shaped obstacle test. Moreover, the wall following approach with fuzzy speed is noted to be much slower, especially in the cluttered environment test and in the V-shaped test. This slower speed for the robot with fuzzy speed control is referred to for one or more of the following reasons: Most trap enclosures force the robot to keep moving with a high rate of rotations and wide heading angles, specifically before detecting the local minima. This reduces the speed of the robot because the heading angle is a strong factor that controls the speed in the proposed fuzzy speed controller. The difference in speed rates of the robot can be observed in Figure 15, where thick and dark paths mean a low speed. The path in the figure indicates the locations of the steps taken by the robot. The robot's behavior is not the same when the speed is controlled, and that is because the step length chosen by the fuzzy system varies from one step to another. As an example, let us consider a robot with a constant step length. The robot in step S at location L1 measures the distance between itself and the nearest obstacle, say D1. Depending on the value of D1, the robot determines the next step's heading angle as in the proposed original navigation system [9]. On the other hand, if the speed is controlled, the same step S will navigate the robot to another location L2, because the step length is different. In this case, the robot measures a different distance D2 to the nearest obstacle. As a result, the robot's decision of the heading angle will also be different. This variance in behavior improved the performance, on average, of the reflected virtual target approach over the local path backtracking approach in general.

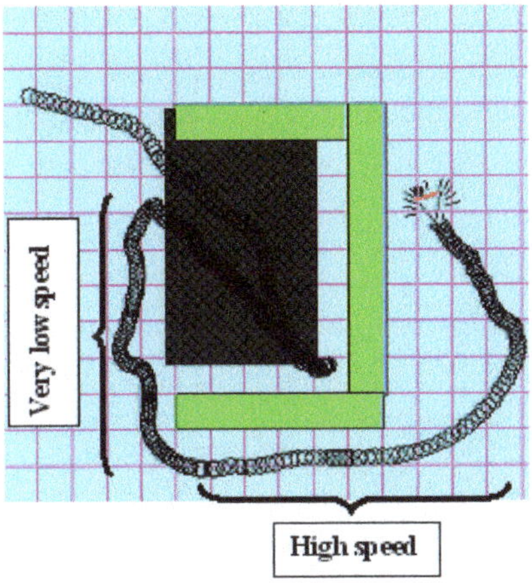

Figure 15. The difference in speed rates when the speed is controlled by the proposed fuzzy speed controller.

In the wall-following approach, the robot is much slower because it keeps moving along with the obstacles' walls. This means that during the wall-following behavior, the robot is very near to obstacles, which also reduces its speed dramatically, as noted in Figure 16.

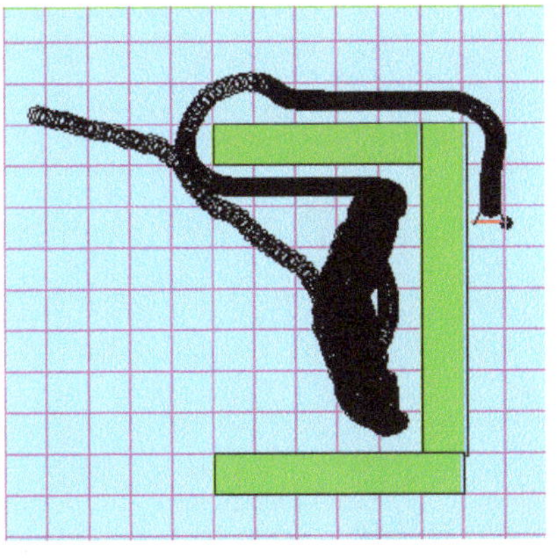

Figure 16. Wall-following approach reaches very low-speed rates when the robot follows the walls.

In environments that include local minima situations, most of the robot steps are either inside the deadlock enclosure or outside but very close. This means that the factor ρ should be higher in these environments, and the speed should be much lower. The wall-following approach's performance was significantly improved when the speed was controlled with the proposed fuzzy system for the E-shaped, C-shaped, double U-shaped, and W-shaped obstacles test cases. This is because, in these test cases, the robot did not follow the whole environment perimeter's wall, as shown in Figure 17 when set side by side with Figure 9f.

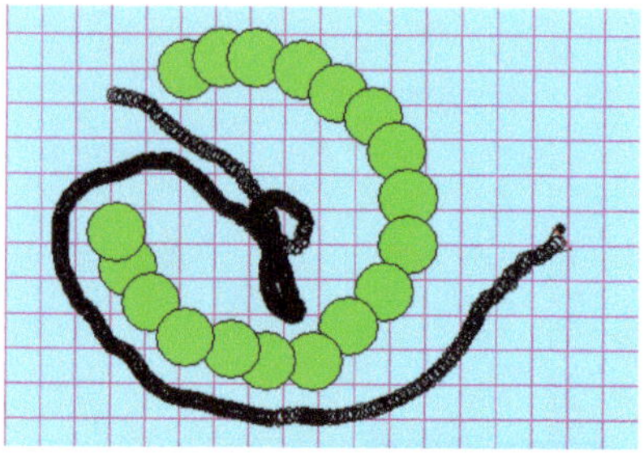

Figure 17. Wall-following performance in the C-shaped obstacle test case.

6. Conclusions and Future Work

6.1. Limitations and Future Work

When the opportunities offered by IoT technologies are exploited fully in an industrial setup, this creates a rich and ubiquitous environment. IoT devices mounted on various objects ranging from vehicles, humans, robots, goods, to moving and still objects all contribute with data that would transform a typical warehouse setting into a well-connected and dynamic smart space. As such, when robot's path planning algorithms are empowered by the data contributed by IoT devices, the IoT system, such as the location of objects, presence of obstacles, and the dynamic changes in the environment, enormous opportunities and improvement to path planning and obstacle avoidance, will arise. The solutions proposed in this work enhance robots' navigation in device-to-device decentralized setups. While the results were verified using simulation work, performance results in real-world experimental scenarios remain to be validated. Future work is planned to enable the full extent of IoT incorporation into the robots' navigational systems. The plan is to set up a number of IoT-enabled robots in a shared environment with other IoT devices and obstacles. We will then attempt to further optimize the proposed solution and make better use of the data supplied by the IoT system. In addition, the local minima problem can be addressed using more than one robot with deep learning solutions, especially in real-life domain space. Robotic co-workers may present and work collaboratively with their peers (human workers) concurrently. Having more robots in a smart factory space could earnestly require a robust system to handle the unpredicted movement of multiple robots in a closed area, such as in a factory.

6.2. Conclusions

In robot navigation systems in IoT, there are main goals that must be achieved, including seeking the goal, avoiding obstacles, and avoiding local minima. In this work, five approaches to address the local minima problem are proposed. Their implications and opportunities in the context of IoT and Industry 5.0 were also highlighted. The reliability of the five proposed approaches and the achieved enhancement in performance was validated by comparing the five approaches to the popular wall-following approach. The results show a significant advantage and improvement in performance in the four test cases conducted in this work. A significant improvement in performance was reported specifically in the cases where the robot encountered W-shaped, C-shaped, and double U-shaped obstacles. Additionally, in cluttered environments, the proposed approaches minimized the distance and time required by the robot to reach its destination by eight times when compared to the traditional path planning approaches. Overall, the results showed that the robot using any of the five proposed approaches requires fewer steps to reach the destination, ranging from 59 to 73 m on average across varied obstacle forms, as opposed to the wall-following strategy, which requires an average of 732 m. On average, the robot with a constant speed and reflected virtual target approach takes 103 s to complete the tasks, which is the greatest performance among the other approaches, whereas the identical robot with a wall-following approach takes 907 s. Using a fuzzy-speed robot, the duration for the wall-following approach is greatly reduced, from 907 to 507 s, while the reflected virtual target, random target, and global path backtracking may only need up to 20%t of that time; 94, 96, and 97 s, respectively. This can be attributed to the fact that the robot using the wall-following approach keeps moving along with the obstacles' walls in order to avoid the local minima.

Author Contributions: Conceptualization, Y.T. and I.H.-M.; methodology, Y.T., I.H.-M., B.A., and O.D.; software, Y.T., I.H.-M., and M.M.; validation, Y.T., O.D., B.A., and M.M.; formal analysis, I.H.-M., B.A., O.D., and M.M.; investigation, Y.T., I.H.-M., and M.E.; resources, Y.T., O.D., M.M., B.A., and N.A.; data curation, Y.T. and I.H.-M.; writing—original draft preparation, Y.T., I.H.-M., B.A., O.D., and M.M.; writing—review and editing, B.A., O.D., M.M., M.E., and N.A.; visualization, Y.T., I.H.-M., B.A., O.D., and M.M.; supervision, Y.T.; project administration, O.D., M.M., and B.A. All authors have read and agreed to the published version of the manuscript.

Funding: This research received no external funding.

Data Availability Statement: Authors can confirm that all relevant data are included in the article.

Conflicts of Interest: The authors declare no conflict of interest.

References

1. Harapanahalli, S.; Mahony, N.O.; Hernandez, G.V.; Campbell, S.; Riordan, D.; Walsh, J. Autonomous Navigation of mobile robots in factory environment. *Procedia Manuf.* **2019**, *38*, 1524–1531. [CrossRef]
2. Nahavandi, S. Industry 5.0—A Human-Centric Solution. *Sustainability* **2019**, *11*, 4371. [CrossRef]
3. Kaiser, M.S.; Al Mamun, S.; Mahmud, M.; Tania, M.H. Healthcare Robots to Combat COVID-19. In *COVID-19: Prediction, Decision-Making, and its Impacts*; Santosh, K., Joshi, A., Eds.; Springer: Singapore, 2021; pp. 83–97. [CrossRef]
4. Fang, B.; Guo, X.; Wang, Z.; Li, Y.; Elhoseny, M.; Yuan, X. Collaborative task assignment of interconnected, affective robots towards autonomous healthcare assistant. *Future Gener. Comput. Syst.* **2019**, *92*, 241–251. [CrossRef]
5. Farid, F.; Elkhodr, M.; Sabrina, F.; Ahamed, F.; Gide, E. A smart biometric identity management framework for personalised IoT and cloud computing-based healthcare services. *Sensors* **2021**, *21*, 552. [CrossRef]
6. Demirhan, M.; Premachandra, C. Development of an Automated Camera-Based Drone Landing System. *IEEE Access* **2020**, *8*, 202111–202121. [CrossRef]
7. Premachandra, C.; Tamaki, M. A Hybrid Camera System for High-Resolutionization of Target Objects in Omnidirectional Images. *IEEE Sens. J.* **2021**, *21*, 10752–10760. [CrossRef]
8. Tashtoush, Y.; Haj-Mahmoud, I. Fuzzy Speed Controller for Mobile Robots Navigation in Unknown Static Environments. In Proceedings of the International Conference on Digital Information Processing, Beijing, China, 21–22 April 2013; p. 139.
9. Al-Jarrah, O.M.; Tashtoush, Y.M. Mobile robot navigation using fuzzy logic. *Intell. Autom. Soft Comput.* **2007**, *13*, 211–228. [CrossRef]
10. Boldrer, M.; Andreetto, M.; Divan, S.; Palopoli, L.; Fontanelli, D. Socially-Aware Reactive Obstacle Avoidance Strategy Based on Limit Cycle. *IEEE Robot. Autom. Lett.* **2020**, *5*, 3251–3258. [CrossRef]
11. Grover, J.S.; Liu, C.; Sycara, K. Deadlock Analysis and Resolution for Multi-Robot Systems. In *Proceedings of the International Workshop on the Algorithmic Foundations of Robotics XIV*; Springer International Publishing: Berlin/Heidelberg, Germany, 2021; pp. 294–312.
12. Mohanty, P.K.; Kodapurath, A.A.; Singh, R.K. A Hybrid Artificial Immune System for Mobile Robot Navigation in Unknown Environments. *Iran. J. Sci. Technol. Trans. Electr. Eng.* **2020**, *44*, 1619–1631. [CrossRef]
13. Wahab, M.N.A.; Nefti-Meziani, S.; Atyabi, A. A comparative review on mobile robot path planning: Classical or meta-heuristic methods? *Annu. Rev. Control* **2020**, *50*, 233–252. [CrossRef]
14. Abiyev, R.; Ibrahim, D.; Erin, B. Navigation of mobile robots in the presence of obstacles. *Adv. Eng. Softw.* **2010**, *41*, 1179–1186. [CrossRef]
15. Xie, Y.; Zhang, X.; Meng, W.; Zheng, S.; Jiang, L.; Meng, J.; Wang, S. Coupled fractional-order sliding mode control and obstacle avoidance of a four-wheeled steerable mobile robot. *ISA Trans.* **2021**, *108*, 282–294. [CrossRef] [PubMed]
16. Cuevas, F.; Castillo, O.; Cortés-Antonio, P. Omnidirectional four wheel mobile robot control with a type-2 fuzzy logic behavior-based strategy. In *Intuitionistic and Type-2 Fuzzy Logic Enhancements in Neural and Optimization Algorithms: Theory and Applications*; Springer: Berlin/Heidelberg, Germany, 2020; pp. 49–62.
17. Zhu, A.; Yang, S.X. Neurofuzzy-Based Approach to Mobile Robot Navigation in Unknown Environments. *IEEE Trans. Syst. Man Cybern. Part C (Appl. Rev.)* **2007**, *37*, 610–621. [CrossRef]
18. Back, S.; Cho, G.; Oh, J.; Tran, X.T.; Oh, H. Autonomous UAV Trail Navigation with Obstacle Avoidance Using Deep Neural Networks. *J. Intell. Robot. Syst.* **2020**, *100*, 1195–1211. [CrossRef]
19. Ben Jabeur, C.; Seddik, H. Design of a PID optimized neural networks and PD fuzzy logic controllers for a two-wheeled mobile robot. *Asian J. Control* **2021**, *23*, 23–41. [CrossRef]

20. Yang, S.; Li, T.; Shi, Q.; Bai, W.; Wu, Y. Artificial Potential-Based Formation Control with Collision and Obstacle Avoidance for Second-order Multi-Agent Systems. In Proceedings of the 2020 7th International Conference on Information, Cybernetics, and Computational Social Systems (ICCSS), Guangzhou, China, 13–15 November 2020; IEEE: Guangzhou, China, 2020; pp. 58–63. [CrossRef]
21. Receveur, J.B.; Victor, S.; Melchior, P. Autonomous car decision making and trajectory tracking based on genetic algorithms and fractional potential fields. *Intell. Serv. Robot.* **2020**, *13*, 315–330. [CrossRef]
22. Li, C.; Cui, G.; Lu, H. The design of an obstacle avoiding trajectory in unknown environment using potential fields. In Proceedings of the 2010 IEEE International Conference on Information and Automation, Harbin, China, 20–23 June 2010; pp. 2050–2054.
23. Csiszar, A.; Drust, M.; Dietz, T.; Verl, A.; Brisan, C. Dynamic and interactive path planning and collision avoidance for an industrial robot using artificial potential field based method. In *Mechatronics*; Springer: Berlin/Heidelberg, Germany, 2011; pp. 413–421.
24. Li, G.; Tamura, Y.; Yamashita, A.; Asama, H. Effective improved artificial potential field-based regression search method for autonomous mobile robot path planning. *Int. J. Mechatron. Autom.* **2013**, *3*, 141. [CrossRef]
25. Javaid, M.; Haleem, A.; Singh, R.P.; Suman, R. Substantial Capabilities of Robotics in Enhancing Industry 4.0 implementation. *Cogn. Robot.* **2021**, *1*, 58–75. [CrossRef]
26. Barosz, P.; GoÅ,da, G.; Kampa, A. Efficiency Analysis of Manufacturing Line with Industrial Robots and Human Operators. *Appl. Sci.* **2020**, *10*, 2862. [CrossRef]
27. Carrasco, P.; Cuesta, F.; Caballero, R.; Perez-Grau, F.J.; Viguria, A. Multi-Sensor Fusion for Aerial Robots in Industrial GNSS-Denied Environments. *Appl. Sci.* **2021**, *11*, 3921. [CrossRef]
28. Rogowski, A.; Skrobek, P. Object Identification for Task-Oriented Communication with Industrial Robots. *Sensors* **2020**, *20*, 1773. [CrossRef] [PubMed]
29. Le, A.V.; Nhan, N.H.K.; Mohan, R.E. Evolutionary Algorithm-Based Complete Coverage Path Planning for Tetriamond Tiling Robots. *Sensors* **2020**, *20*, 445. [CrossRef] [PubMed]
30. Krishna, K.M.; Kalra, P.K. Solving the local minima problem for a mobile robot by classification of spatio-temporal sensory sequences. *J. Robot. Syst.* **2000**, *17*, 549–564. [CrossRef]
31. Nurmaini, S. Intelligent navigation in unstructured environment by using memory-based reasoning in embedded mobile robot. *Eur. J. Sci. Res.* **2012**, *72*, 228–244.
32. Ordonez, C.; Collins, E.G.; Selekwa, M.F.; Dunlap, D.D. The virtual wall approach to limit cycle avoidance for unmanned ground vehicles. *Robot. Auton. Syst.* **2008**, *56*, 645–657. [CrossRef]
33. Um, D.; Ryu, D.; Kang, S. A framework for unknown environment manipulator motion planning via model based realtime rehearsal. In *Intelligent Autonomous Systems 12*; Springer: Berlin/Heidelberg, Germany, 2013; pp. 623–631.
34. Taylor, K.; LaValle, S.M. I-Bug: An intensity-based bug algorithm. In Proceedings of the 2009 IEEE International Conference on Robotics and Automation, Kobe, Japan, 12–17 May 2009; pp. 3981–3986.
35. Lumelsky, V.; Stepanov, A. Dynamic path planning for a mobile automaton with limited information on the environment. *IEEE Trans. Autom. Control* **1986**, *31*, 1058–1063. [CrossRef]
36. Sanchez, G.M.; Giovanini, L.L. Autonomous navigation with deadlock detection and avoidance. In *Sociedad Iberoamericana de Inteligencia Artificial*; CONICET: Buenos Aires, Argentina, 2014.

computers

Article

Affecting Young Children's Knowledge, Attitudes, and Behaviors for Ultraviolet Radiation Protection through the Internet of Things: A Quasi-Experimental Study

Sotiroula Theodosi and Iolie Nicolaidou *

Department of Communication and Internet Studies, Cyprus University of Technology, P.O. Box 50329, Limassol 3603, Cyprus; soti.theodosi@gmail.com
* Correspondence: iolie.nicolaidou@cut.ac.cy; Tel.: +357-2500-2105

Abstract: Prolonged exposure to ultraviolet (UV) radiation is linked to skin cancer. Children are more vulnerable to UV harmful effects compared to adults. Children's active involvement in using Internet of Things (IoT) devices to collect and analyze real-time UV radiation data is suggested to increase their awareness of UV protection. This quasi-experimental pre-test post-test control group study implemented light sensors in a STEM inquiry-based learning environment focusing on UV radiation and protection in primary education. This exploratory, small-scale study investigated the effect of a STEM environment implementing IoT devices on 6th graders' knowledge, attitudes, and behaviors about UV radiation and protection. Participants were 31 primary school students. Experimental group participants (n = 15) attended four eighty-minute inquiry-based lessons on UV radiation and protection and used sensors to measure and analyze UV radiation in their school. Data sources included questionnaires on UV knowledge, attitudes, and behaviors administered pre- and post-intervention. Statistically significant learning gains were found only for the experimental group ($t_{14} = -3.64$, $p = 0.003$). A statistically significant positive behavioral change was reported for experimental group participants six weeks post-intervention. The study adds empirical evidence suggesting the value of real-time data-driven approaches implementing IoT devices to positively influence students' knowledge and behaviors related to socio-scientific problems affecting their health.

Keywords: Internet of Things (IoT); ultraviolet (UV) radiation protection; primary education; STEM intervention; inquiry learning

Citation: Theodosi, S.; Nicolaidou, I. Affecting Young Children's Knowledge, Attitudes, and Behaviors for Ultraviolet Radiation Protection through the Internet of Things: A Quasi-Experimental Study. *Computers* **2021**, *10*, 137. https://doi.org/10.3390/computers10110137

Academic Editor: Sergio Correia

Received: 16 September 2021
Accepted: 21 October 2021
Published: 25 October 2021

1. Introduction

Overexposure to solar ultraviolet (UV) radiation is a risk for public health [1]. Excessive and prolonged exposure to UV radiation can lead to adverse effects, including some eye diseases, diseases associated with vitamin D insufficiency [2], premature aging, sunburn, and skin cancers [3]. As the incidence of skin cancer is increasing rapidly, it has become one of the greatest threats to public health and has created a substantial economic burden for skin cancer treatment, particularly in countries such as New Zealand, Australia, the USA, UK, and Germany [3,4], as well as in Europe [5]. Children are considered a high-risk population group [6] and are more vulnerable to the sun's harmful effects compared to adults, as their skin is thinner and more sensitive, and even a short time outdoors in the midday sun can result in serious burns [6,7]. Skin cancer is steadily increasing in prevalence, faster than any other cancer, disproportionately affecting a growing youth population [8]. Melanoma cases, perhaps the most aggressive type of cancer, are increasing, and so are other skin cancer types at increasingly younger ages [6,8,9].

The most proactive and effective way of preventing skin cancer is through education, increasing awareness of the dangers of UV radiation exposure, and promoting sun protection practices [3,6]. Researchers agree that the dangers of skin cancer can be ameliorated through prevention efforts, especially those targeted at children [10]. From a young age, children need to be educated about the sun's harmful effects on the skin and how best to protect themselves [7,11].

1.1. Related Work on Measuring Young Children's Sun-Related Attitudes and Knowledge

Most of the studies that focused on measuring young children's sun-related attitudes and knowledge followed a survey design that was based on a one-time administration of a questionnaire to a large sample of students, sometimes following a non-technologically supported intervention [7,10,12,13]. For example, using a structured questionnaire, in Aquilina et al.'s study, 965 Maltese secondary school students were surveyed concerning their sun-related attitudes and knowledge. The study demonstrated a high level of sun awareness among Maltese secondary school students [7]. In the study by Aquilina et al. (2004), despite the high knowledge scores and relatively positive attitudes, the rate of deliberate suntanning was still high, especially for girls. Their study showed that although necessary, knowledge is not enough for a change in attitude and even less so for a sustained behavior change. In the study by Wright et al., a randomly selected sample of 488 children from 27 primary schools in New Zealand was surveyed regarding their sun-related knowledge, attitudes, and behaviors. Wright et al. (2008) found that although knowledge increased with school year level, there was a decline in sun-protective attitudes and behaviors [13]. Saridi et al. (2012) [6] recorded habits and attitudes of 2163 primary school students in Greece regarding sun-protection measures and pointed out the necessity of health education programs for students and parents/teachers alike to raise awareness about everyday sun protection.

1.2. Related Work on Sun-Safe Prevention Interventions in Primary Education

Although skin cancer is an easily preventable disease, self-directed prevention behaviors in children are difficult to achieve [10]. Behaviorally-based intervention strategies are needed to facilitate the transition from knowledge to a change in attitude and then to a change in behavior [7]. As most children spend most of the peak hours for ultraviolet radiation at school, school instruction on sun-safe behaviors and attitudes is a popular primary prevention method. Environmental interventions (including providing places with shade and free sunscreen) were found to have a promising role in skin cancer prevention interventions among children and adolescents [5]. Some studies that used behavioral interventions in primary and secondary schools also had positive results [10,12,14], reporting significant improvements for knowledge and attitudes [14] and knowledge, attitudes, and behaviors [12]. Hart and DeMarco (2008) [10] found that primary prevention interventions have seen the most success in elementary schools rather than secondary schools, particularly when interventions used multiunit extended instruction over time, to augment students' knowledge of sun-safe behaviors and attitudes toward skin protection and to encourage students to practice more sun-protective behaviors [10].

1.3. State-of-the-Art Analysis on Using Internet of Things to Measure Ultraviolet Radiation

Previous prevention behavioral interventions [5,10,12,14] were not technologically supported. Contemporary homes are filled with digital technologies, and children are exposed to them almost since birth, initiating their first digital experiences at very early ages. In a world where computers, the Internet, and mobile devices such as smartphones and tablets are widely used for e-health [15], this trend is expected to become stronger. Our future has been envisioned around the concept of the IoT (Internet of Things) [16], a global computing infrastructure of trillions of connected devices that permeate the world we live in [17]. The IoT refers to a connection between sensors and actuators implanted into physical objects [18,19]. Ubiquitous computing and the Internet of things (IoT) are turning into everyday household technology at an ever-increasing pace, for example, in the form of connected toys [20] and in the form of sensors that can be used for data collection, processing and visualization [21]. By embedding sensors, an IoT network can collect rich sensor data that reflect real-time environment conditions [21], such as UV radiation levels in a specific environment.

An extensive number of studies were conducted to measure personal solar UV exposure in various settings [2], but these studies were not applied in education. Some recent studies focused on how UV radiation can be quantified and measured in real-time through the IoT [22] for measuring and controlling environmental pollution [23] and for positively affecting public health [1]. To facilitate the monitoring of solar UV intensity and cumulative dose, various UV sensors were developed in the past few decades, and many are commercially available [3], but they were not used for educational purposes.

1.4. The Necessity of This Study

IoT in education is a new conceptual paradigm that is still in its starting phase with a potential benefit and impact in the learning process that has not yet been realized [24–26]. Although there were several contributions on the inclusion of IoT into the education domain [27,28], there is still a lack of consolidated and coherent view on this topic [21], and there are several socio-technological challenges associated with it [29]. There is also a minimal number of educational applications that take advantage of IoT's potential in the learning process and are appropriate for lower grades of formal education, as the majority of applications targets higher education [27,30,31] or secondary education [32]. Despite opportunities for data collection, processing and visualization made possible with IoT devices for STEM activities [21,27,30], very few studies have focused on how the potential of IoT can be realized in primary education [27]. For example, only 11% (10 papers) of studies that employed IoT in education and were included in the literature review of [21] targeted elementary school students. Sensors and IoT data collection technologies allow students to be engaged in actual research in the way scientists are, and this helps students build an understanding of science concepts [29]. However, no studies were identified that used the IoT to influence young children's UV protection knowledge, attitudes and behaviors.

The study responds to the pressing need for increasing children's awareness of the dangers of UV radiation exposure by implementing a data-driven approach and IoT to influence primary school children's sun-related knowledge, attitudes, and behaviors; thus addressing a research gap identified in the literature. The study describes an inquiry-based activity sequence in which students had active involvement using IoT devices to analyze real-time UV radiation data collected through light sensors. This was expected to engage learners in an inquiry on their surroundings in a data-rich physical and digital school environment. A quasi-experimental pre-test post-test control group research design was used to evaluate the effectiveness of the proposed approach in increasing students' UV radiation knowledge and in positively affecting their attitudes concerning the need for UV radiation protection, a socio-scientific issue affecting their health.

2. Materials and Methods

2.1. Research Questions

This study focused on designing a STEM inquiry learning environment about UV radiation and protection by implementing IoT devices. It was enacted in formal primary education to answer the following research questions:

RQ1: To what extent does a STEM environment implementing IoT devices affect 6th-grade students' knowledge regarding UV radiation and protection?

RQ2: To what extent does a STEM environment implementing IoT devices affect 6th-grade students' attitudes, behaviors, and behavioral change regarding UV radiation protection?

2.2. Implementation of IoT in Primary Education

An activity sequence incorporating IoT devices consisting of four eighty-minute ultraviolet radiation and protection courses was designed and developed based on inquiry-based learning (IBL) principles [33]. With respect to technical equipment, students used commercially available sensors to measure ultraviolet radiation in their schoolyard (Pasco Wireless Light sensors), which were wirelessly connected to tablets, allowing students to collect data in real-time and observe and process these data through the Sparkvue software. Digital PASPORT Sensors contain an analog-to-digital converter and are automatically recognized by PASCO software when connected through an interface. Most of these sensors contain multiple sensing elements within one casing, which enables them to collect different measurements using a single interface port. The Wireless Light Sensor is a coin cell battery-powered wireless sensor that connects to a computer or tablet device through Bluetooth. The sensor measures light through two apertures. The Spot Light Aperture measures red-green-blue (RGB), and white. The Ambient Light Aperture measures illuminance (measured in lux or lumens per square meter), Photosynthetically Active Radiation (PAR) in sunlight, and solar irradiance (in watts per square meter). The Ambient Light Aperture also measures UVA (ultra-violet A) and UVB (ultra-violet B) allowing ultra-violet index (UVI) to be calculated. PASCO Data Collection Software displays and analyzes the measurements from the sensor. The software also supports remote data logging for long-term experiments. Since each sensor has a unique Device ID number, more than one can be connected to a computer or tablet at the same time [34]. The proposed sensors can be integrated into the primary school curriculum circumventing some of the technical challenges described in the literature [29].

With respect to the activity sequence, during the first lesson, students studied multi-modal sources focusing on the question "Why is the sun dangerous?". Students worked individually due to social distancing protocols enforced during the COVID-19 pandemic and discussed their findings. They then experimented with UV beads focusing on the question "Can we make UV light visible?" and suggested making UV beads bracelets to measure the extent of UV radiation exposure in their schoolyard. Students used their UV bracelets outdoors and made observations regarding UV radiation.

In the second lesson, students addressed the following problem: "The headmaster needs to know during which hours are UV index levels high to take some precautionary measures for students. How can you help?" Students studied the UV index and realized that precise measurement of UV radiation levels using appropriate instruments is needed. Students worked in pairs, placed light sensors in the schoolyard and in the football field, and measured UV radiation levels in real-time during the lesson and for the next two weeks.

In the third lesson, students interpreted sensor-collected data regarding UV radiation levels at their school and suggested products they considered as sun-protective (e.g., sunglasses, hats, and sunscreen) and desirable actions while outdoors to ensure protection from UV radiation (e.g., playing in the shade).

In the fourth lesson, students used sensors to examine suggested sun-protective products to determine their level of protection. They worked outdoors in groups of four and later discussed their findings in class.

2.3. Data Sources

Data sources were the following: (1) a test for assessing students' knowledge regarding UV radiation and protection [7,35], (2) the Greek translation of the questionnaire measuring attitudes [7] and behaviors [36] concerning UV radiation and protection, and (3) the Greek translation of the RASP-B questionnaire for categorizing participants in the stages of the transtheoretical model of behavioral change concerning UV radiation and protection [36].

The test assessing students' knowledge of UV radiation and protection consisted of 20 closed-ended (right or wrong) statements consistent with course objectives and validated in other studies assessing learning gains regarding UV radiation and protection [7,35]. The closed-ended questions were grouped into three categories: (a) questions referring to the risk of excessive sun exposure (questions 1–5), (b) questions relating to ultraviolet radiation (questions 6–12), and (c) questions concerning ways of protection against excessive sun exposure (questions 13–20). The researchers added the open-ended question "What do you know about UV radiation?" to eliminate the possibility of randomly selected answers to the closed-ended questions (Appendix A). The test's content validity was evaluated by an expert in Science Education and pilot-tested with a small number of students.

A modified version of the Aquilina et al. (2004) [7] questionnaire was used to investigate students' attitudes towards UV radiation and protection. The modification refers to changing the dichotomous "Yes/No" scale to a 5-point Likert scale indicating frequency of behavior or level of agreement for greater interpretation capacity. A modified version of the Borschmann and Cottrell (2009) [36] questionnaire was used to investigate students' behaviors towards UV radiation and protection. The original questionnaire included 14 statements, of which five were eliminated due to their inconsistency with course objectives. Participants were asked to choose an option from a 5-point Likert scale ranging from 1 (never) to 5 (always) (Appendix B).

The RASP-B questionnaire was used [36] to record any change in participants' stage of behavior towards UV protection. It included 12 statements, which allocate participants in the first three stages of the behavioral change model: a. Pre-contemplation (questions 1, 5, 10, and 12), b. Contemplation (questions 3, 4, 8 and 9) and c. Action (questions 2, 6, 7 and 11). Participants were asked to choose one option on a 5-point Likert scale ranging from 1 (I completely disagree) to 5 (I completely agree) (Appendix C). The RASP-B questionnaire appears to have satisfactory psychometric properties, with the Cronbach's alpha ranging from 0.67 to 0.76. Both questionnaires were translated into Greek and then back-translated for accuracy.

2.4. Research Design

The study followed a quasi-experimental pre-test–post-test control group design. Delayed post-tests were also administered to experimental group students six weeks after the intervention for examining knowledge, attitudes, behaviors, and stage of behavior preservation. Primary school-age children are generally more responsive to efforts to increase sun-safe behaviors and improve attitudes toward skin cancer prevention than are adolescents [10]. Therefore, the study focused on upper primary school students. A primary school was chosen through convenience sampling. There were two sixth-grade classes in the school selected for the study. One class (to which the first author had access) was purposively selected as the experimental group; hence the other served as a control group.

2.5. Participants

Fifteen 6th-grade primary school students (eight boys and seven girls) were included in the experimental group and participated in four eighty-minute courses on UV radiation and protection during March and April 2021. Sixteen 6th-grade primary school students (nine boys and seven girls) constituted the control group and did not receive any instruction regarding UV radiation and protection.

2.6. Research Ethics

All participating students and their parents were informed in writing about the study's objectives and gave their informed consent for inclusion before they participated in the study. The study was conducted in accordance with the Declaration of Helsinki and it adhered to the ethical standards of the American Psychological Association and General Data Protection Regulation guidelines. The study protocol was approved by the Ethics Committee of the Center of Educational Research and Evaluation of the country in reference (approval code 7.15.01.25.8.1/10) on 4 August 2020. All participants, and their parents' written consent was ensured.

2.7. Data Analysis

Descriptive statistics (mean, SD) and inferential statistical tests (paired samples and independent samples *t*-tests) were used for data analysis in SPSS 26 for answering both research questions. To verify the integrity of the outcomes, a comparative analysis using R, a programming language and environment for statistical computing and graphics, was performed for statistically significant results. Closed-ended questions of the knowledge test for RQ1 were scored dichotomously by allocating 1 point for any correct answer and 0 points for any wrong answer [37]. Phenomenography was applied to score the open-ended question [38] based on the emerging coding scheme shown in Table 1.

Table 1. Coding scheme for the open-ended question "What do you know about UV radiation?

Points	Rationale	Example
0	No answer Wrong answer Uses the question to answer	*"I don't know" (Student 9)* *"I do not know exactly, but I think it is the light from the sun" (Student 15)* *"I know it's a type of radiation (Student 7)"*
1	Refers to the sun as the source of the UV radiation Refers to some characteristics showing the severity of UV radiation, its effects on human health and the necessity to adopt protective behaviors.	*"I know it's harmful to the human body"* *"[...] we mustn't spend too much time in the sun, because we will get sunburned. If we want to stay (in the sun) we must take precautions" (Student 8)*
2	Refers to specific characteristics of UV radiation such as name, source, types Refers to severity as one of its characteristics and gives an example Refers to effects on human health and gives examples of precautionary measures	*"It has three types UVA, UVB and UVC. UVC can't pass through the ozone layer. The other two can, and in specific hours and seasons they can be very dangerous for humans" (Student 12)* *"Ultraviolet radiation (UV) is divided in UVA, UVB and UVC and it's dangerous when in high level. We need to be protected with proper precautionary measures so that it won't harm our skin." (Student 13)* *"UV radiation comes from the sun and can cause severe damage to eyes and skin." (Student 14)*

A second researcher coded all students' answers. Inter-rated reliability among the two coders was calculated at 81.81%, with four disagreements resolved among the two researchers.

For RQ2 referring to attitudes, the options *never/completely disagree* and *rarely/disagree* were considered as desired attitudes (for example: the response "completely disagree" in a statement such as the following: "I think getting sunburned occasionally does not do any harm" is considered a desired attitude). Desired attitudes were evaluated with one point; otherwise, they were evaluated with zero points. For investigating participants' behaviors, options *sometimes, usually,* and *always* were evaluated with one point since they were classified as desirable behaviors. For example: the response "usually" in a statement such as the following: "How often do you wear a hat when outside in the sun" is considered a desired behavior. In any different case, behaviors were evaluated with zero points. The total score of students' attitudes and behaviors was computed, respectively. An α level of 0.05 was set a priori for all statistical analyses.

Regarding participants' classification of the stages of behavioral change, the model suggested by [36] was used. First, the options were scored as follows: −2 points (completely disagree), −1 point (disagree), 0 points (no opinion), +1 point (agree), +2 points (completely agree). Each stage had four corresponding statements. The total grade of each participant for each stage of behavior was calculated. The total score ranged from −8 points to +8 points, and the participant was classified to the stage with the highest score.

3. Results

3.1. Research Question 1

All 31 students were pre-tested regarding their knowledge regarding UV radiation and protection prior to the intervention to ensure group equivalence (Table 2). There were no significant differences concerning the initial scores achieved by the experimental group (Mean = 14.67; SD = 2.12) compared with the scores achieved by the control group

(Mean = 12.06; SD = 3.06), when an independent samples *t*-test was performed ($t_{29} = 2.7$, $p = 0.253$).

Table 2. Students' knowledge of UV radiation and protection before and after the intervention for the experimental and control group.

	n	Pre		Post		Delayed	
		Mean	SD	Mean	SD	Mean	SD
Experimental group	15	14.67	2.12	17.73 **	2.63	18.00 **	2.12
Control group	16	12.06	3.06	11.56	3.98		

** Significant result with $p < 0.01$.

After ensuring group equivalence, experimental students' pre and post-intervention scores were compared by performing a paired samples *t*-test. Experimental group participants showed statistically significant learning gains, in contrast with control group participants who did not (Table 2). The post-intervention experimental group students' performance significantly increased from 14.67 (SD 2.12) to 17.73 (SD 2.63) ($t_{14} = -3.64$, $p = 0.003$). This result was confirmed using R. Control group students' test performance decreased slightly but not significantly from pre- to post-intervention (Table 2).

For investigating knowledge preservation, a test was administered six weeks after the intervention in the experimental group. As shown in Table 2, learning gains significantly increased from 14.67 (SD 2.12) pre-intervention to 18.00 (SD 2.12) ($t_{14} = -4.152$, $p = 0.001$) six weeks after the intervention. This result was confirmed using R.

3.2. Research Question 2

All 31 students' attitudes regarding UV radiation and protection were measured prior to the intervention. An independent samples *t*-test was performed to establish group equivalence. Both experimental (Mean 3.71; SD 0.52) and control group participants (Mean 3.52; SD 0.69) held similar attitudes concerning UV radiation and protection, which were considerably high for both groups. No significant differences were noted ($t_{29} = 0.844$; $p = 0.406$), therefore groups were considered equivalent regarding their attitudes towards UV radiation and protection prior to the intervention (Table 3).

Table 3. Attitudes on UV protection pre- and post-intervention for the experimental and control group.

	n	Pre		Post	
		Mean	SD	Mean	SD
Experimental group	15	3.71	0.52	3.77	0.65
Control group	16	3.52	0.69	3.70	0.54

For comparing students' pre- and post-intervention attitudes for both experimental and control group participants, a paired samples *t*-test was performed. Experimental group participants' attitudes regarding UV radiation and protection were slightly but not significantly increased from 3.71 (SD 0.52) to 3.77 (SD 0.65). A similar pattern was revealed for control group participants, whose attitudes had a slight but non-significant increase from 3.52 (SD 0.69) to 3.70 (SD 0.54). As experimental group students' attitudes were considerably high from the beginning (approximating 4 out of 5), they were not measured again in the delayed post-test.

All 31 students' behaviors regarding UV protection were investigated prior to the intervention for group equivalence to be established. An independent samples *t*-test was performed. Experimental (Mean 3.15; SD 0.70) and control group participants (Mean 3.00; SD 0.58) reported similar behaviors concerning UV protection. No significant differences were noted ($t_{29} = 0.656$; $p = 0.427$), therefore the two groups were considered equivalent regarding their behaviors towards UV protection prior to the intervention (Table 4).

Table 4. Students' behaviors on UV protection in the experimental and control group before and after intervention.

	n	Pre		Post		Delayed	
		Mean	**SD**	**Mean**	**SD**	**Mean**	**SD**
Experimental group	15	3.15	0.70	3.25	0.71	3.52 *	0.71
Control group	16	3.00	0.58	2.88	0.87		

* Significant result with $p < 0.05$.

Experimental group participants' behavior regarding UV protection was positively differentiated from pre- to post-intervention. Pre-intervention behaviors held by experimental group participants were considered neutral at first (Mean 3.15; SD 0.70), moved to more desired ones immediately after the intervention (Mean 3.25; SD 0.71), and continued to be significantly and positively differentiated (Mean 3.52; SD 0.71) six weeks after the intervention ($t_{14} = -2.46$; $p = 0.027$), a finding confirmed using R. On the contrary, behaviors adopted from control group participants were negatively differentiated from 3.00 (SD 0.58) to 2.88 (0.87).

As far as UV protection behavioral change is concerned, an improvement regarding the desired behavior was recorded for experimental group participants. The latter moved from hierarchically lower levels of behavior concerning UV protection to higher ones immediately after the intervention, based on the Transtheoretical Model of Behavioral Change (TMBC) (Table 5). Table 5 shows an increase in students categorized in the higher level of behavior measured, the action stage, from one student post-intervention to four students in the delayed post-test, a finding that is not observed in the control group, where we see a decrease in students categorized in the action stage from pre to post.

Table 5. Classification of participants' behaviors in the experimental and control group based on Transtheoretical Model of Behavioral Change (TMBC).

Stages of Behavior	**Frequencies of Desired Behaviors**				
	Pre	**Post**	**Delayed**	**Pre**	**Post**
	Experimental Group			**Control Group**	
Pre-contemplation stage	3	1	4	13	8
Contemplation stage	10	11	7	0	6
Action stage	1	2	4	3	2

Specifically, experimental group participants positively differentiated their post-intervention behavior pattern concerning UV protection for the statements "How often do you wear clothes that cover most of your body to avoid the sun?" (from 4 to 10 participants), "How often do you wear a hat when exposed to the sun?" (from 9 to 12) and "How often do you sit in the shade when you are in the beach?" (from 12 to 14). Six weeks after, participants' behavior regarding UV protection remained positive. Specifically, a positive differentiation was presented in six out of seven statements of sun-protective behaviors referring to "How often will you go sunbathing between 11:00 to 4:00?" (from 5 to 7), "How often do you wear a hat when exposed to the sun?" (from 12 to 13), "How often do you sit in the shade when you are in the beach?" (from 12 to 13), "How often do you wear sunglasses?" (from 10 to 14), "How often do you wear sunscreen?" (from 14 to 15), "How often do you stay in the shade to avoid the sun?" (from 10 to 13). Despite the improvement noted in behaviors adopted from experimental group participants, six weeks after the intervention, some participants regressed to lower stages of behavioral change (Table 5).

As for the control group, their post behaviors differentiated, with the number of participants in the contemplation stage rising from zero to six and the number of participants in the pre-contemplation stage declining from thirteen to eight. Positive changes observed in control group participants' self-reported attitudes and desired behavioral stages can potentially be attributed to the diffusion of treatment intervention [39]. The latter was inevitable in this study as students of the two groups shared the school spaces where the UV sensors were placed. They can also be attributed to control group participants' increased motivation for positive self-presentation [40] and an effort to present themselves in the most positive manner possible to outperform experimental group participants.

4. Discussion

The present study aimed to investigate the extent to which a STEM inquiry-based learning environment implementing IoT devices can affect sixth-grade students' knowledge, attitudes, and behaviors about UV radiation and protection. Regarding RQ1, findings demonstrated that experimental group students' learning gains were improved significantly by the end of the intervention, consistent with previous work reporting that non-technologically supported interventions regarding UV radiation and protection contributed to knowledge acquisition [7,13,14,35,41]. Contrary to previous studies that did not measure learning gains using delayed post-tests, our study showed that students' learning gains were preserved six weeks post-intervention. This may indicate that interventions incorporating IoT devices that students can use to collect data and solve socio-scientific problems that affect them personally are more likely to result in long-term learning effects.

Concerning RQ2, both experimental and control group students' attitudes towards UV radiation and protection remained positive from pre- to post-intervention with a slight but non-significant improvement by the end of the intervention. Moreover, experimental students' behaviors regarding UV protection were positive pre- and post-intervention and significantly improved six weeks after the intervention took place while control group students' behaviors deteriorated. Studies have shown that young children's attitudes on complex socio-scientific issues are difficult to change [42]. Studies have also shown the limitations of short-duration efforts. Although they can improve children's knowledge and, in some cases, improve attitudes and sun-protection behavior immediately after the program, their influence is likely to be short-lived [7]. Contrary to previous research findings, in our study, a delayed post-test showed that the effect of the intervention had been retained over time, at least with respect to self-reported behavioral change and intended practices for the upcoming summer.

The differentiating factor in the behavioral intervention described in our study as compared with the ones described in the literature thus far [10,12,14] was the integration of IoT devices in the curriculum to positively affect children's UV protection knowledge, attitudes and behaviors, through a real-time data-driven inquiry learning approach. The effectiveness of the described STEM intervention may potentially be attributed to several design characteristics. One of them is the authentic problem students were asked to solve that made them realize the need for real-time and accurate measurement of UV radiation levels through sensors. Another design characteristic refers to providing the ability to visualize, analyze and interpret meaningful data that students collected themselves using IoT devices to make informed decisions as to the extent to which specific products provide UV radiation protection. The use of IoT for real-time UV radiation data collection and analysis contributed to improving students' knowledge on UV radiation protection, in enhancing meaningful, active learning in an authentic environment that afforded experimentation to find solutions to real-life problems, as indicated by [27] and in positively influencing children's UV protection behaviors.

IoT is set to transform the education domain in many ways in the near future [21]. This exploratory study resulted in an empirically validated IoT intervention to increase young children's UV protection knowledge and positively affect desired attitudes and behaviors. The study contributed to the lack of published studies involving the effective integration of IoT in primary education STEM topics, which was introduced at a pilot stage in the country where the study took place in 2019. It increased our understanding of how innovative technologies combined with a data-driven approach can positively influence students' knowledge, attitudes, and behaviors related to socio-scientific problems affecting their health.

4.1. Limitations

Convenience sampling was used rather than random sampling, which would have been preferred to increase the generalizability of the study's findings. In addition, the first author held a dual role as the intervention designer/teacher and researcher. The study was exploratory, used a small sample, and was based solely on quantitative data. Due to the small sample of the study, the use of advanced Artificial Intelligence/Machine Learning techniques in order to provide more reliable results was not feasible. The use of a larger sample and the addition of qualitative data in the form of in-depth student interviews or observations conducted by parents or teachers concerning children's attitudes and behaviors regarding UV protection would strengthen the study.

The fact that a quasi-experimental design was used, in which the control group did not participate in an innovative intervention and did not receive any instruction, threatened the study's internal validity; a limitation commonly reported in quasi-experiments in the social sciences [43–45]. As previously mentioned, the data collected by control group participants might have been affected by factors such as "secondary" treatment infusion [39] and positive self-presentation [40]. Moreover, without objective measures of children's practice, we had to rely on their self-reports, which are susceptible to memory errors and social desirability tendencies, a problem also reported by [14].

4.2. Future Work

A number of devices and innovative methodologies have recently been promoted, advocating the benefits of monitoring personal exposure patterns to solar radiation. These innovations are beneficial to the community and researchers as tools for monitoring sun exposure behavior [2]. For example, Wu et al. (2019) [1] developed an IoT application in which when high UV is detected, it will notify users by pushing notifications to their mobile phone so that the users will take some appropriate actions [1]. Using objective measures such as wearables to monitor actual exposure has potential in future studies that will aim to measure the effect of behavioral interventions more accurately and systematically [5,46].

Our future work will first identify reliable and practical UV sensors for personal UV exposure monitoring [3] that are appropriate for use by school children, such as UV monitor bracelets as wearable devices [47]. It will then use wearable devices measuring UV radiation to triangulate children's self-reported data and measure the effect of IoT-based behavioral interventions on children's attitudes and behaviors more accurately and over time in future longitudinal studies with larger sample sizes. Future work will also empower students by involving them in a co-design process of IoT devices for inquiry learning, as Kusmin (2019) [27] suggested.

Author Contributions: Conceptualization, S.T.; Methodology, S.T. and I.N.; Software, S.T., Validation, S.T. and I.N.; Formal Analysis, S.T. and I.N.; Investigation, S.T.; Resources, S.T.; Data Curation, S.T.; Writing—Original Draft Preparation, I.N.; Writing—Review and Editing, S.T. and I.N.; Visualization, S.T.; Supervision, I.N.; Project Administration, S.T. All authors have read and agreed to the published version of the manuscript.

Funding: This research received no external funding.

Institutional Review Board Statement: The study was conducted in accordance with the Declaration of Helsinki and it adhered to the ethical standards of the American Psychological Association and General Data Protection Regulation guidelines. The study protocol was approved by the Ethics Committee of the Center of Educational Research and Evaluation of the country in reference (approval code 7.15.01.25.8.1/10) on 4 August 2020.

Informed Consent Statement: All participating students and their parents were informed in writing about the study's objectives and gave their informed consent for inclusion before they participated in the study. All participants, and their parents' written consent was ensured.

Data Availability Statement: The data presented in this study are available on request from the corresponding author. The data are not publicly available due to privacy reasons.

Acknowledgments: The authors would like to thank the students who voluntarily participated in the study.

Conflicts of Interest: The authors declare no conflict of interest.

Appendix A. Test Assessing Students' Knowledge on UV Radiation and Protection

1. What do you know about UV radiation?

2. Read the statements carefully and circle the correct answer.

	A1. Knowledge of the risk of excessive sun exposure		
1.	Too much sun can harm people.	TRUE	FALSE
2.	Too much sun exposure can cause freckles.	TRUE	FALSE
3.	Too much sun exposure can cause wrinkles on the skin when you grow older.	TRUE	FALSE
4.	The sun is bad for your skin only when you get sunburnt.	TRUE	FALSE
5.	A suntan is a sign of being healthy.	TRUE	FALSE
	A2. Knowledge of UV radiation		
6.	UV rays are sun rays.	TRUE	FALSE
7.	UV rays are always dangerous for people.	TRUE	FALSE
8.	Ultraviolet radiation can harm your skin and eyes, only if you spend too much time in the sun.	TRUE	FALSE
9.	UV level is high all day long.	TRUE	FALSE
10.	UV rays helps in synthesizing vitamin D.	TRUE	FALSE
11.	UV index uses colors to show how danger UV rays are.	TRUE	FALSE
12.	UV rays can cause suntan and sunburn to the skin.	TRUE	FALSE
	A3. Knowledge of protection against excessive sun exposure		
13.	The sun is strongest and more harmful between 11 a.m. and 4:00 p.m.	TRUE	FALSE
14.	The ozone layer protects the earth from too much ultraviolet radiation.	TRUE	FALSE
15.	You cannot get too much sun, on a cloudy day.	TRUE	FALSE
16.	You do not need to protect your skin with sunscreen when under a beach umbrella.	TRUE	FALSE
17.	If you apply sunscreen, there is no need to wear a hat and shirt when in the sun.	TRUE	FALSE
18.	One application of sunscreen protects your skin for at least 4 h.	TRUE	FALSE
19.	The sun can harm your eyes, and you should wear sunglasses when out in the sun.	TRUE	FALSE
20.	Sunscreen with Sun Protection Factor less than 15 is not enough to protect you.	TRUE	FALSE

Appendix B. Questionnaire Assessing Students' Attitudes and Behaviors towards UV Radiation and Protection

PART A: In this part, statements concerning people's attitudes towards UV protection are included.					
Instructions: In the next table, circle the number that represents you the most for each one of the following statements. For your answers use the scale 1 to 5. **1 = Never 2 = Rarely 3 = Sometimes 4 = Usually 5 = Always**					
A1. Attitudes towards UV protection					
1. I am shy to apply sunscreen in front of my friends.	1	2	3	4	5
2. Wearing a shirt at the beach does not look cool.	1	2	3	4	5
3. Covering up in the sun is a hassle.	1	2	3	4	5
4. I worry that the sun may give me freckles.	1	2	3	4	5
1 = Completely disagree 2 = Disagree 3 = Neither agree nor disagree 4 = Agree 5 = Completely agree					
A2. Attitudes towards UV protection					
1. My parents do not protect themselves from the sun, so I do not feel that I need to be careful myself.	1	2	3	4	5
2. I think getting sunburnt occasionally does not do any harm.	1	2	3	4	5
3. I do not feel that I should protect myself from the sun because skin cancer can never happen to me.	1	2	3	4	5
PART B: In this part, questions and statements regarding sun-protective actions are included.					
Instructions: In the following table, circle the number showing how often each of the following statements represents you. For your answers, use the following scale of 1 to 5. **1 = Never 2 = Rarely 3 = Sometimes 4 = Usually 5 = Always**					
B1. Behavior/Perceived severity					
1. In the past, how often did you go sunbathing at the beach between 11:00 a.m. and 4:00 p.m.?	1	2	3	4	5
2. How often do you wear clothes that cover most of your body (hands and legs included) to avoid the sun?	1	2	3	4	5
3. How often do you wear a hat when outside in the sun?	1	2	3	4	5
4. How often do you sit in the shade at the beach?	1	2	3	4	5
5. How often do you wear sunglasses?	1	2	3	4	5
6. How often do you use sunscreen?	1	2	3	4	5
7. How often do you stay in the shade to avoid the sun?	1	2	3	4	5
Instructions: In the following table, circle Yes or No.					
B2. Behavior					
1. Last summer, did you get reddish skin or a sunburn on your face or body?	YES			NO	
2. Think of a summer day. How many hours do you spend exposed to the sun? Put a tick in a box.					
a. 0 h	d. three to four hours				
b. less than an hour	e. more than five hours				
c. one to two hours					

Appendix C. RASP-B Questionnaire Assessing Students' Stage of Behavior towards UV Radiation Protection

RASP-B Questionnaire					
Statements referring to sun exposure and their peoples' beliefs about it are included in this questionnaire. Read the instructions carefully and answer all the following statements.					
Instructions: In the following table, circle the number representing your level of agreement for each of the following statements. For your answers, use the following scale of 1 to 5.					
1 = Completely disagree 2 = Disagree 3 = Neither agree nor disagree 4 = Agree 5 = Completely agree					
1. I do not think I spend too much time exposed to the sun.	1	2	3	4	5
2. I am trying to spend less time in the sun than I used to.	1	2	3	4	5
3. I enjoy spending time in the sun, but sometimes I spend too much time in the sun.	1	2	3	4	5
4. Sometimes I think I should spend less time in the sun.	1	2	3	4	5
5. It's a waste of time thinking about how much time I spend in the sun.	1	2	3	4	5
6. I have just recently changed my sun exposure habits.	1	2	3	4	5
7. Anyone can talk about wanting to do something reducing their sun exposure, but I am actually doing something about it.	1	2	3	4	5
8. I am at a stage where I should think about spending less time in the sun.	1	2	3	4	5
9. The amount of time I spend in the sun is a problem sometimes.	1	2	3	4	5
10. There is no need for me to think about changing my sun exposure habits.	1	2	3	4	5
11. I am actually changing my sun exposure habits right now.	1	2	3	4	5
12. Spending less time in the sun would be pointless for me.	1	2	3	4	5

References

1. Wu, F.; Wu, T.; Yuce, M.R. An Internet-of-Things (IoT) Network System for Connected Safety and Health Monitoring Applications. *Sensors* **2019**, *19*, 21. [CrossRef] [PubMed]
2. Downs, N.J.; Parisi, A.V.; Butler, H.; Rawlings, A.; Elrahoumi, R.S. An Inexpensive High-Temporal Resolution Electronic Sun Journal for Monitoring Personal Day to Day Sun Exposure Patterns. *Front. Public Health* **2017**, *5*, 310. [CrossRef] [PubMed]
3. Huang, X.; Chalmers, A.N. Review of Wearable and Portable Sensors for Monitoring Personal Solar UV Exposure. *Ann. Biomed. Eng.* **2021**, *49*, 964–978. [CrossRef] [PubMed]
4. Gordon, L.G.; Rowell, D. Health system costs of skin cancer and cost-effectiveness of skin cancer prevention and screening: A systematic review. *Eur. J. Cancer Prev.* **2015**, *24*, 141–149. [CrossRef]
5. Thoonen, K.; van Osch, L.; de Vries, H.; Jongen, S.; Schneider, F. Are Environmental Interventions Targeting Skin Cancer Prevention among Children and Adolescents Effective? A Systematic Review. *Int. J. Environ. Res. Public Health* **2020**, *17*, 529. [CrossRef] [PubMed]
6. Saridi, M.; Toska, A.; Rekleiti, M.; Wozniak, G.; Liachopoulou, A.; Kalokairinou, A.; Souliotis, K.; Birbas, K. Sun-Protection Habits of Primary Students in a Coastal Area of Greece. *J. Ski. Cancer* **2012**, *2012*, e629652. [CrossRef] [PubMed]
7. Aquilina, S.; Gauci, A.A.; Ellul, M.; Scerri, L. Sun awareness in Maltese secondary school students. *J. Eur. Acad. Dermatol. Venereol.* **2004**, *18*, 670–675. [CrossRef]
8. Robertson, F.M.-L.; Fitzgerald, L. Skin cancer in the youth population of the United Kingdom. *J. Cancer Policy* **2017**, *12*, 67–71. [CrossRef]
9. Siegel, R.; DeSantis, C.; Virgo, K.; Stein, K.; Mariotto, A.; Smith, T.; Cooper, D.; Gansler, T.; Lerro, C.; Fedewa, S.; et al. Cancer treatment and survivorship statistics, 2012. *CA A Cancer J. Clin.* **2012**, *62*, 220–241. [CrossRef] [PubMed]
10. Hart, K.M.; DeMarco, R.F. Primary Prevention of Skin Cancer in Children and Adolescents: A Review of the Literature. *J. Pediatric Oncol. Nurs.* **2008**, *25*, 67–78. [CrossRef] [PubMed]
11. Sümen, A.; Öncel, S. Development of sun protection behaviors in preschoolers: A systematic review. *Turkderm* **2018**, *52*, 56–63. [CrossRef]
12. Kouzes, E. Sun Smart Schools Nevada: Increasing Knowledge among School Children about Ultraviolet Radiation. *Prev. Chronic Dis.* **2017**, *14*, E125. [CrossRef] [PubMed]
13. Wright, C.; Reeder, A.I.; Gray, A.; Cox, B. Child sun protection: Sun-related attitudes mediate the association between children's knowledge and behaviours. *J. Paediatr. Child Health* **2008**, *44*, 692–698. [CrossRef] [PubMed]

14. Geller, A.C.; Rutsch, L.; Kenausis, K.; Selzer, P.; Zhang, Z. Can an hour or two of sun protection education keep the sunburn away? Evaluation of the Environmental Protection Agency's Sunwise School Program. *Environ. Health* **2003**, *2*, 13. [CrossRef] [PubMed]
15. Tozzi, F.; Nicolaidou, I.; Galani, A.; Antoniades, A. eHealth interventions for anxiety management targeting young children and adolescents: Exploratory review. *JMIR Pediatrics Parent.* **2018**, *1*, e7248. [CrossRef] [PubMed]
16. Brito, R.; Dias, P.; Oliveira, G. Young children, digital media and smart toys: How perceptions shape adoption and domestication. *Br. J. Educ. Technol.* **2018**, *49*, 807–820. [CrossRef]
17. Kortuem, G.; Bandara, A.K.; Smith, N.; Richards, M.; Petre, M. Educating the Internet-of-Things Generation. *Computer* **2013**, *46*, 53–61. [CrossRef]
18. Akhter, F.; Siddiquei, H.R.; Alahi, M.E.E.; Mukhopadhyay, S.C. Recent Advancement of the Sensors for Monitoring the Water Quality Parameters in Smart Fisheries Farming. *Computers* **2021**, *10*, 26. [CrossRef]
19. Teo, T.; Unwin, S.; Scherer, R.; Gardiner, V. Initial teacher training for twenty-first century skills in the Fourth Industrial Revolution (IR 4.0): A scoping review. *Comput. Educ.* **2021**, *170*, 104223. [CrossRef]
20. Mertala, P. Young children's perceptions of ubiquitous computing and the Internet of Things. *Br. J. Educ. Technol.* **2020**, *51*, 84–102. [CrossRef]
21. Kassab, M.; DeFranco, J.; Laplante, P. A systematic literature review on Internet of things in education: Benefits and challenges. *J. Comput. Assist. Learn.* **2020**, *36*, 115–127. [CrossRef]
22. Baena-Navarro, R.; Torres-Hoyos, F.; Vergara-Villadiego, J.; Gómez, A.P. Quantification of gamma and UV radiation using the Internet of Things. *Proc. ISSSD* **2019**, *3*, 244–258.
23. Saha, H.N.; Auddy, S.; Chatterjee, A.; Pal, S.; Pandey, S.; Singh, R.; Singh, R.; Sharan, P.; Banerjee, S.; Ghosh, D.; et al. Pollution control using Internet of Things (IoT). In Proceedings of the 2017 8th Annual Industrial Automation and Electromechanical Engineering Conference (IEMECON), Bangkok, Thailand, 16–18 August 2017; pp. 65–68. [CrossRef]
24. Abdel-Basset, M.; Manogaran, G.; Mohamed, M.; Rushdy, E. Internet of things in smart education environment: Supportive framework in the decision-making process. *Concurr. Comput. Pract. Exp.* **2019**, *31*, e4515. [CrossRef]
25. Burd, B.; Barker, L.; Divitini, M.; Perez FA, F.; Russell, I.; Siever, B.; Tudor, L. Courses, Content, and Tools for Internet of Things in Computer Science Education. In Proceedings of the 2017 ITiCSE Conference on Working Group Reports, Bologna, Italy, 3–5 July 2017; pp. 125–139. [CrossRef]
26. Elsaadany, A.; Soliman, M. Experimental Evaluation of Internet of Things in the Educational Environment. *Int. J. Eng. Pedagog.* **2017**, *7*, 50–60. [CrossRef]
27. Kusmin, M. Co-Designing the Kits of IoT Devices for Inquiry-Based Learning in STEM. *Technologies* **2019**, *7*, 16. [CrossRef]
28. Rahim, R.; Sudarsana, I.K.; Napitupulu, D.; Listyorini, T.; Kurniasih, N.; Manurung, R.; Sallu, S. Humidity and Temperature Prototype for Education with Internet of Things. *Int. J. Pure Appl. Math.* **2018**, *119*, 2487–2490.
29. Alfoudari, A.M.; Durugbo, C.M.; Aldhmour, F.M. Understanding socio-technological challenges of smart classrooms using a systematic review. *Comput. Educ.* **2021**, *173*, 104282. [CrossRef]
30. Kusmin, M.; Saar, M.; Laanpere, M.; Rodríguez-Triana, M.J. Work in progress—Smart schoolhouse as a data-driven inquiry learning space for the next generation of engineers. In Proceedings of the 2017 IEEE Global Engineering Education Conference (EDUCON), Athens, Greece, 25–28 April 2017; pp. 1667–1670. [CrossRef]
31. Lai, Y.-H.; Chen, S.-Y.; Lai, C.-F.; Chang, Y.-C.; Su, Y.-S. Study on enhancing AIoT computational thinking skills by plot image-based VR. *Interact. Learn. Environ.* **2021**, *29*, 482–495. [CrossRef]
32. Stojanović, D.; Bogdanović, Z.; Petrović, L.; Mitrović, S.; Labus, A. Empowering learning process in secondary education using pervasive technologies. *Interact. Learn. Environ.* **2020**, 1–14. [CrossRef]
33. Pedaste, M.; Mäeots, M.; Siiman, L.A.; de Jong, T.; van Riesen SA, N.; Kamp, E.T.; Manoli, C.C.; Zacharia, Z.C.; Tsourlidaki, E. Phases of inquiry-based learning: Definitions and the inquiry cycle. *Educ. Res. Rev.* **2015**, *14*, 47–61. [CrossRef]
34. Wireless Light Sensor Pack. PS-3338. Available online: https://www.pasco.com/products/sensors/wireless/ps-3338 (accessed on 6 October 2021).
35. Gooderham, M.J.; Guenther, L. Sun and the Skin: Evaluation of a Sun Awareness Program for Elementary School Students. *J. Cutan. Med. Surg.* **1999**, *3*, 230–235. [CrossRef] [PubMed]
36. Borschmann, R.D.; Cottrell, D. Developing the readiness to alter sun-protective behaviour questionnaire (RASP-B). *Cancer Epidemiol.* **2009**, *33*, 451–462. [CrossRef] [PubMed]
37. Chen, G.; Shen, J.; Barth-Cohen, L.; Jiang, S.; Huang, X.; Eltoukhy, M. Assessing elementary students' computational thinking in everyday reasoning and robotics programming. *Comput. Educ.* **2017**, *109*, 162–175. [CrossRef]
38. Nicolaidou, I.; Kyza, E.A.; Terzian, F.; Hadjichambis, A.; Kafouris, D. A framework for scaffolding students' assessment of the credibility of evidence. *J. Res. Sci. Teach.* **2011**, *48*, 711–744. [CrossRef]
39. Gundersen, K.; Svartdal, F. Diffusion of treatment interventions: Exploration of 'secondary' treatment diffusion. *Psychol. Crime Law* **2010**, *16*, 233–249. [CrossRef]
40. Christensen, L.B.; Waraczynski, M.A. *Experimental Methodology*; Allyn and Bacon: Boston, MA, USA, 1988.
41. Grant-Petersson, J.; Dietrich, A.J.; Sox, C.H.; Winchell, C.W.; Stevens, M.M. Promoting Sun Protection in Elementary Schools and Child Care Settings: The SunSafe Project. *J. Sch. Health* **1999**, *69*, 100–106. [CrossRef]

42. Herodotou, C.; Kyza, E.A.; Nicolaidou, I.; Hadjichambis, A.; Kafouris, D.; Terzian, F. The Development and Validation of the GMOAS, an Instrument Measuring Secondary School Students' Attitudes Towards Genetically Modified Organisms. *Int. J. Sci. Educ. Part B* **2012**, *2*, 131–147. [CrossRef]
43. Nicolaidou, I.; Venizelou, A. Improving Children's E-Safety Skills through an Interactive Learning Environment: A Quasi-Experimental Study. *Multimodal Technol. Interact.* **2020**, *4*, 10. [CrossRef]
44. Nicolaidou, I.; Stavrou, E.; Leonidou, G. Building Primary-School Children's Resilience through a Web-Based Interactive Learning Environment: Quasi-Experimental Pre-Post Study. *JMIR Pediatrics Parent.* **2021**, *4*, e27958. [CrossRef]
45. Nicolaidou, I.; Pissas, P.; Boglou, D. Comparing Immersive Virtual Reality to Mobile Applications in Foreign Language Learning in Higher Education: A Quasi-Experiment. *Interact. Learn. Environ.* **2021**, 1–15. [CrossRef]
46. Nagelhout, E.S.; Lensink, R.; Zhu, A.; Parsons, B.G.; Jensen, J.D.; Wu, Y.P. The Feasibility and Acceptability of Using a Wearable UV Radiation Exposure Monitoring Device in Adults and Children: Cross-Sectional Questionnaire Study. *JMIR Dermatol.* **2020**, *3*, e15711. [CrossRef] [PubMed]
47. Wei, J. How Wearables Intersect with the Cloud and the Internet of Things: Considerations for the developers of wearables. *IEEE Consum. Electron. Mag.* **2014**, *3*, 53–56. [CrossRef]

MDPI AG
Grosspeteranlage 5
4052 Basel
Switzerland
Tel.: +41 61 683 77 34

Computers Editorial Office
E-mail: computers@mdpi.com
www.mdpi.com/journal/computers

www.ingramcontent.com/pod-product-compliance
Lightning Source LLC
Chambersburg PA
CBHW061516210926
51850CB00008B/1939
9783725816224